Muhammed Hasnain Kabir Nayeem

DEVELOPMENT OF AL-BASED METAL MATRIX COMPOSITES

Muhammed Hasnain Kabir Nayeem

DEVELOPMENT OF AL-BASED METAL MATRIX COMPOSITES

Reinforced with Hybrid Nanoparticles and its Machinability Test

LAP LAMBERT Academic Publishing

Imprint
Any brand names and product names mentioned in this book are subject to trademark, brand or patent protection and are trademarks or registered trademarks of their respective holders. The use of brand names, product names, common names, trade names, product descriptions etc. even without a particular marking in this work is in no way to be construed to mean that such names may be regarded as unrestricted in respect of trademark and brand protection legislation and could thus be used by anyone.

Cover image: www.ingimage.com

Publisher:
LAP LAMBERT Academic Publishing
is a trademark of
Dodo Books Indian Ocean Ltd. and OmniScriptum S.R.L publishing group

120 High Road, East Finchley, London, N2 9ED, United Kingdom
Str. Armeneasca 28/1, office 1, Chisinau MD-2012, Republic of Moldova, Europe
Managing Directors: Ieva Konstantinova, Victoria Ursu
info@omniscriptum.com

Printed at: see last page
ISBN: 978-620-8-44251-4

DEVELOPMENT OF AL-BASED METAL MATRIX COMPOSITES REINFORCED WITH HYBRID NANO PARTICLES AND ITS MACHINABILITY TEST

By

Muhammed Hasnain Kabir Nayeem

DEDICATION

Dedicated to my parents for supporting and encouraging me to believe in myself.

ACKNOWLEDGMENT

At first, the author expresses his heartiest thanks to the Almighty for giving the patience and potential to complete the thesis work. I want to express my appreciation to all the people who have given their hearts whelming full support in preparing and completing the study. I would like to express my utmost gratitude to my supervisor Dr. Md. Sayem Hossain Bhuiyan, for his inspiring advice, patient guidance, constructive suggestions, and enthusiastic supervision. The research area that Dr. Md. Sayem Hossain Bhuiyan led me to was interesting and the skills I learned from it will help me a lot in my future career. Without his tremendous support and encouragement, this thesis would not have materialized. Many thanks also go to the Grad-coordinator, Aeronautical Engineering Department, MIST who provided the lab facilities whenever required.

The author wishes to express his sincere gratitude to Air Cdre Md Maksudul Alam, BUP, psc, Senior Instructor and Head, Department of Aeronautical Engineering, MIST for his utmost support and guidance regarding excellent research environment.

The author also wishes to convey his thanks and gratitude to Dr. Shahida Begum, Professor, Department of Mechanical Engineering, MIST, and Dr. Md. Afsar Ali, Professor, Department of Mechanical Engineering, BUET for serving their committee. Their comments, suggestions and approvals have undoubtedly enhanced the quality of the work.

The author also is grateful to Md. Touhidur Rahman, Department of Aeronautical Engineering, MIST, helped me a lot through numerous tips during different stages of the thesis work. The author wish to express his heartfelt gratitude to the Metal Casting Department staff members, BITAC for helping in the fabrication and production of the job. All staff of the Machine tools lab, Measurement, and Quality control lab, Material processing lab, MIST for their help in any respect during the research work.

An honorable mention goes to my beloved parents who inspired and fully supported me in every trial that came my way. Lastly, I thank all those who supported me in any respect during the research work.

ABSTRACT

Development of Al-based Metal Matrix Composites Reinforced with Hybrid Nano Particles and its Machinability Test

Aluminum is a significant material for all type of manufacturing industries due to its wide range of alloys and composites. Despite the fact that aluminum alloys have been used in a wide range of industries due to their excellent and diverse functional characteristics, composite materials can be modified to provide specific mechanical and tribological properties. An aluminum alloy with lesser hardness and tensile strength has been strengthened dramatically by adding ceramic reinforcements.

This study aims at formulate and develop hybrid particle reinforced Al-metal matrix composites using stir casting method, a novel fabrication process in our country to fabricate metal matrix composite. Also, followed by investigation of its the physical, mechanical, morphological and machinability characteristics of the developed metal matrix composites. Three unique compositions of composites have been obtained by varying the wt. % amount of carbon nanotube, alumina and silicon carbide particulate reinforcements respectively 1%, 2.5% and 2.5%. Various mechanical properties, essentially the tensile strength, flexural strength, hardness, impact resistance; and physical properties like porosity, and density were tested; and a morphological and machinability study has been carried out for investigating the performance of the newly developed composites. The study showed that there was an 128.57% increment of tensile strength, an 7.1349% increment of hardness, an 45% increment of impact resistance, and a 0.8301% reduction of density by adding the particulate reinforcements in aluminum metal matrix composite.

The morphological analysis demonstrated a more homogenous dispersion of reinforcement particles in the composite, which indicated the effectiveness of the stir-casting fabrication method. From the optimized machining parameters, it was evident that higher cutting speed and feed rate can be obtained by introducing multiple particulate reinforcement in the metal matrix, which eventually increased the productivity, efficiency and product quality of the developed composites.

সার সংক্ষেপ

Development of Al-based Metal Matrix Composites Reinforced with Hybrid Nano Particles and its Machinability Test

অ্যালুমিনিয়াম তার বিস্তৃত ধাতু এবং কম্পোজিটগুলির কারণে সমস্ত ধরণের উত্পাদন শিল্পের জন্য একটি উল্লেখযোগ্য উপাদান। অ্যালুমিনিয়াম অ্যালয়গুলি তাদের চমৎকার এবং বৈচিত্র্যময় কার্যকরী বৈশিষ্ট্যগুলির কারণে বিস্তৃত শিল্পে ব্যবহার করা হয়েছে তা সত্ত্বেও, নির্দিষ্ট যান্ত্রিক এবং ট্রাইবোলজিক্যাল ক্ষমতা প্রদানের জন্য যৌগিক উপকরণগুলি পরিবর্তন করা যেতে পারে। কম কঠোরতা এবং প্রসার্য শক্তি সহ একটি অ্যালুমিনিয়াম খাদকে সিরামিক শক্তিবৃদ্ধি যোগ করে নাটকীয়ভাবে শক্তিশালী করা হয়েছে।

এই অধ্যয়নের লক্ষ্য হল হাইব্রিড পার্টিকেল রিইনফোর্সড আল-মেটাল ম্যাট্রিক্স কম্পোজিট তৈরি করা, যা আমাদের দেশে ধাতু ম্যাট্রিক্স কম্পোজিট তৈরি করার জন্য স্টির কাস্টিং পদ্ধতি ব্যবহার করে একটি অভিনব বানোয়াট প্রক্রিয়া। এছাড়াও, উন্নত ধাতব ম্যাট্রিক্স কম্পোজিটগুলির শারীরিক, যান্ত্রিক, রূপগত এবং যন্ত্রের বৈশিষ্ট্যগুলি তদন্ত করুন। কম্পোজিটের তিনটি অনন্য রচনা ওজন এর পরিবর্তন করে প্রাপ্ত হয়েছে। কার্বন ন্যানোটিউব, অ্যালুমিনা এবং সিলিকন কার্বাইড পার্টিকুলেট রিইনফোর্সমেন্টের পরিমাণ যথাক্রমে ১%, ২.৫% এবং ২.৫%। বিভিন্ন যান্ত্রিক বৈশিষ্ট্য, মূলত প্রসার্য শক্তি, নমনীয় শক্তি, কঠোরতা, প্রভাব প্রতিরোধের; এবং পোরোসিটি, এবং ঘনত্বের মতো শারীরিক বৈশিষ্ট্যগুলি পরীক্ষা করা হয়েছিল; এবং নতুন উন্নত কম্পোজিটগুলির কার্যকারিতা তদন্তের জন্য একটি রূপতাত্ত্বিক এবং মেশিনিবিলিটি অধ্যয়ন করা হয়েছে। গবেষণায় দেখা গেছে যে প্রসার্য শক্তির ১২৪.৫৭% বৃদ্ধি, কঠোরতা ৭.১৩৪৯% বৃদ্ধি, প্রভাব প্রতিরোধের ৪৫% বৃদ্ধি, এবং হাইব্রিড ধাতু ম্যাট্রিক্স কম্পোজিটের শক্তি যোগ করে ঘনত্ব ০.৮৩০১% হ্রাস পেয়েছে।

রূপতাত্ত্বিক বিশ্লেষণ কম্পোজিটে শক্তিবৃদ্ধি কণার আরও একজাতীয় বিচ্ছুরণ প্রদর্শন করেছে, যা আলোড়ন-কাস্টিং ফ্যাব্রিকেশন পদ্ধতির কার্যকারিতা নির্দেশ করে। অপ্টিমাইজ করা মেশিনিং পরামিতিগুলি থেকে, এটি স্পষ্ট ছিল যে ধাতব ম্যাট্রিক্সে একাধিক কণা শক্তিবৃদ্ধি প্রবর্তন করে উন্নত কম্পোজিটগুলির উত্পাদনশীলতা, দক্ষতা এবং পণ্যের গুণমান বৃদ্ধি করে উচ্চ কাটিং গতি এবং ফিড রেট পাওয়া যেতে পারে।

Table of Contents

List of Tables

List of Figures

List of Notation

Symbol	**Meaning**
AMMCs	Aluminum metal matrix composites
MMCs	Metal matrix composites
PMCs	Polymer matrix composites
CMCs	Ceramic matrix composites
Al_2O_3	Aluminum oxide (Alumina)
SiC	Silicon carbide
TiB_2	Titanium bromide
TiC	Titanium carbide
MoS_2	Molybdenum disulfide
TiN	Titanium nitride
B_4C	Boron carbide
CNT	Carbon nanotube
Wt.%	Weight percent
ARB	Accumulative roll bending
Vol%	Volume fraction
r/min	Revolution per minute
ANOVA	Analysis of variance
MWCNT	Multi-wall carbon nanotubes
FEM	Finite element method
SA	Simulated annealing
GA	Genetic algorithm
ANN	Artificial neural networking
CAM	Computer aided manufacturing
RSM	Response surface methodology
TOPSIS	Technique for order preference by similarity to ideal solution
ASTM	American society for testing and materials
SEM	Scanning electron microscopy

UTS	Ultimate tensile strength
BHN	Brinell hardness number
D_i	Decision matrix
C	Values of each criteria
X_{ij}	Performance of the ith alternative on the jth criteria
$\overline{X_{ij}}$	Normalized decision matrix
$\bar{V}$	Weighted normalized matrix
W_j	Weights
A_j^+	Ideal best solution
A_j^-	Negative ideal solution
J	J=1, 2, … n and J is associates with benefit criteria
J'	J=1, 2, … n and J is associates with cost/lost criteria
δ_i^+	Euclidean distance of ith alternative from ideal solution
δ_i^-	Euclidean distance of ith alternative from negative ideal solution
C_i^*	Relative closeness
e_j	Entropy of each criteria
D_n	Normalized decision matrix
N_j	Normalized evaluation of alternative
W_i	Significance coefficient
B	Normalized decision matrix
Y_i	Weight of the ith alternative
W_j	Respected weight of the jth criteria
F_d	Delamination factor
D_{hole}	Hole exit diameter
D_{max}	Maximum hole diameter

δ	Taper angle
D_{in}	Hole entrance diameter
D_{out}	Hole exit diameter
L	Specimen thickness
V	Cutting speed
F	Feed
S/N	Signal to noise ratio
Δ_{oi}	Deviation sequence
ξ	Grey relational coefficient
γ	Normalize S/N ratio
$\overline{\gamma_J}$	Grey relational grade
K	Number of performance characteristics
PSI	Preference selection index
A	Initial decision matrix
m	Number of specified criteria
i	Number of alternatives
j	Number of criteria
C_p	Positive criteria
C_n	Negative criteria
b_{ij}	Normalized decision matrix
$\overline{B_J}$	Attained mean value of the jth criteria
m	Total number of criteria
PV_j	Preference variation value of the jth criteria
ϕ_j	Deviation preference value of the jth criteria
ψ_j	Overall performance value
I_i	Preference selection index
R_{ij}	Normalized value of attributes
R_j	Mean of normalized value of attributes j

CHAPTER 1
INTRODUCTION

1.1 Background of the Research

The demands placed on materials for improved performance are so many and varied that no single material can meet all of them. As a result, the traditional idea of integrating diverse elements in an integral-composite material to meet the user's requirements has been revived. Composites have given design engineering a new level of flexibility, effectively requiring the designer-analyst to invent new applications to save weight and expense. Fig. 1.1 shows the increase in the use of composite material in recent years. Metal matrix composites (MMCs) are now widely used in industrial, medical, and construction sectors such as automobiles, airplanes, marine time industries, medical equipment, and construction equipment, because of MMC's desirable features, such as increased strength and wear resistance. Moreover, MMCs are light in weight, have good thermal conductivity, and have a low thermal expansion (Chakrapani and Suryakumari, 2020; Rajasekar et al., 2018).

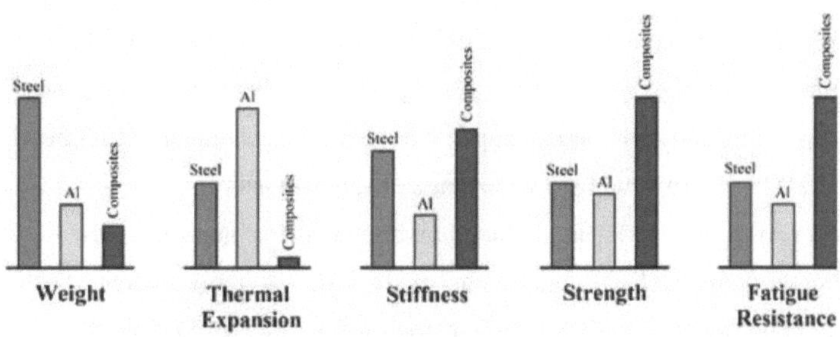

Fig. 1.1: Comparison between traditional material and composite material (Tamizharasan et al., 2019).

Metals that are commonly used as the matrix in the fabrication of MMC are aluminum alloys, magnesium alloys, and titanium alloys (Chakrapani and Suryakumari, 2020; Rajasekar et al. 2018; Kumar et al., 2020; Escaich et al., 2020). On the other side, in the production of metal matrix composite, the material commonly used as reinforcement are aluminum oxide (alumina) (Al_2O_3), silicon carbide (SiC), titanium bromide (TiB_2), titanium carbide (TiC), molybdenum disulfide (MoS_2), titanium nitride

(TiN), boron carbide (B_4C), fly-ash, carbon nano-tube (CNT), graphite and other reinforcing materials (Ravikumar et al. 2021, Yashpal et al. 2020, Kumar et al. 2020, Bhushan 2020, Elumalai et al. 2020, Kumar et al 2017, Rebba and Ramanaiah 2014, Chen and Shih-Fu 2020, Ravindran 2019, Kumar 2017, Devanathan 2020, Sharma et al. 2019, Premnath et al. 2014, Ravikumar et al. 2021).

MMCs have metallic base alloy qualities (such as conductivity, ductility, toughness and hardness) as well as ceramic reinforcement properties (wear resistance, strength, high young modulus, and thermal stability) (Sharma et al. 2019, Verma et al. 2016). As a result of hybridization, improved tensile, shear, and compression strength, improved operating temperature, and reduced wear loss can be achieved.

Aluminum alloys are one of the most used materials in the fabrication of MMCs as matrix material because of their lightweight, strong corrosion resistance, and outstanding electrical and thermal conductivity (Oladijo et al. 2021, Gupta 2018). In the production of composites, several aluminum alloys (e.g. 1xxx, 2xxxx, 3xxxx, 5xxxx, 6xxxx, 7xxxx, and others) are chosen (Kareem et al. 2021). The 6xxx series of aluminum alloys has strong machinability and extrude ability, and is used primarily as a matrix material in the manufacture of AMMCs (Kareem et al. 2021). Because of its low density, high specific strength, high stiffness, better wear resistance, low thermal expansion, and high thermal application zone, thus aluminum metal matrix composite (AMMCs) is suitable for a wide range of applications.

As reinforcing materials in aluminum alloy matrix composites, silicon carbide (SiC), boron carbide (B_4C), aluminum oxide (Al_2O_3), molybdenum disulfide (MoS_2), graphite, titanium carbide (TiC), carbon nano-tube (CNT), fly ash and hybrid nano-particles, etc. were used as reinforcement material (Kareem et al. 2021, Devanathan et al. 2019, Jaykumar and Kohir 2020, Manigandan et al 2015, Ravindran et al. 2019, Callister 2019, Yang et al. 2021, Yeshiye and Gizaw 2021). MMCs contain metallic characteristics of base alloys and ceramics properties of reinforcements (Kareem et al. 2021).

Metal matrix composites can be divided into solid, liquid, or semi-solid form production processes based on the major production operations. Metal matrix composite production methods can be classified as solid state (powder metallurgy), liquid state (such as stir casting, stir squeezed casting), or semi-solid state (such as spray deposition) (Jaykumar and Kohir 2020, Kareem et al. 2021).

The stir casting process was applied to generate the composite specimen within that experiment because it is a simple and cost-effective method of creating large to small-sized MMC components (Kareem et al. 2021).

By adding silicon carbide, alumina, and carbon nano-tube particles to aluminum metal matrix alloy, the major goal of this research is to create hybrid particle-reinforced aluminum metal matrix composite materials with better mechanical and tribological properties than the aluminum alloy metal. Due to its good formability, machinability, wettability, corrosion resistance, and general-purpose alloy, aluminum alloy is chosen as the matrix material for a wide range of structural applications and welded applications, including truck components, railroad, marine applications, agricultural applications, aircraft components, automotive parts, medical equipment, machine parts and more (Ravindran et al. 2019).

The stir casting process is used to manufacture non-reinforced aluminum metal alloy, mono-particle reinforced (Al-alloy + CNT), bi-particle reinforced (Al-alloy + CNT + SiC), and tri-particle reinforced (Al-alloy + CNT + SiC + Al_2O_3) composite materials in this work. The relationship between microstructures and mechanical qualities including hardness, tensile strength, and percentage of elongation, as well as tribological parameters like taper angle, delamination was examined. An optical microscope was used to examine the microstructure of the samples. Optimization of machining parameters was performed to determine the best suitable machining condition.

1.2 Motivation of the Study

Nowadays, manufacturing industries such as automotive, aerospace, and marine industries among others, are focusing on producing lightweight and high-strength components utilizing cost-effective composite materials rather than steel and other heavy materials. Aluminum is the third most prevalent element in the earth's crust and is a light metal with a specific weight of 2.7 gm/cm^3 (about one-third that of steel). Aluminum is used in car rotor drums, clutches, engine blocks, connecting rods, automobile bodies, shafts, gear trains, and other components to minimize inactive weight and energy consumption. In other sense, using heavy metals in automobile components increases the vehicle's overall weight and fuel consumption. Aluminum alloy is widely used for a variety of applications outside of automotive use due to its low density and availability. Though aluminum appeals to users because of its low weight and abundant availability, it still requires improvements in mechanical and tribological qualities such as tensile strength, compressive strength, hardness, and wear resistance to meet the demands of the applications. Furthermore, few studies have focused on the machinability testing of aluminum-based metal matrix composites. This gap in the research serves as one of the motivations for this study, which aims to address this aspect.

Hence, this study focuses on the fabrication of a unique hybrid particle reinforced aluminum matrix composite reinforced with aluminum oxide, silicon carbide, and carbon nanotubes in order to achieve desired mechanical, physical & tribological properties and overcome the existing limitations.

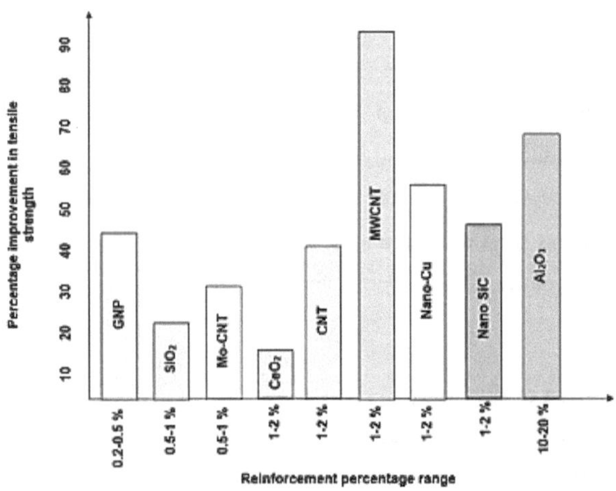

Fig. 1.2: Comparison between percentage improvement in tensile strength with reinforcement percentage range (Tamizharasan et al., 2019).

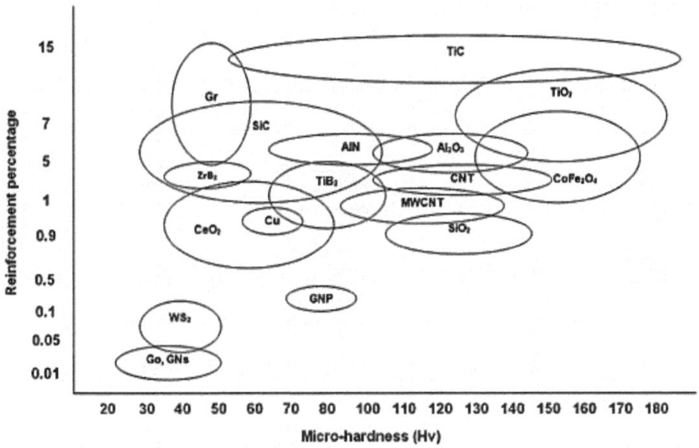

Fig. 1.3: Effect of reinforcing element and micro-hardness value (Bhoi, Singh, and Pratap, 2020).

Fig. 1.2 and Fig. 1.3 provides insight into the improvement in mechanical properties of the aluminum-based metal matrix composite with the introduction of various type of reinforcement. The cost barrier of the automobile or other machinery can be reduced by selecting the most cost-effective reinforced material.

The following are some of the benefits of this research project:

It will be employed as a low-cost and lightweight alternative source of materials. It will reduce the machine's weight, for example, using this hybrid particle reinforced composite reduces the vehicle's weight while also lowering fuel consumption.

The machine component's mechanical and tribological qualities have significantly improved. It will replace the material used in some parts of automobiles, aircraft, and heavy industrial equipment.

Finally, as a teaching aid, and reference source for technological education and future composites research.

1.3 Research Objectives

Main objectives of this work is to fabricate and characterize the hybrid particle reinforced aluminum metal matrix composites, with variation of reinforcements wt.% of SiC, Al_2O_3 and CNT, and optimize the various machining parameters. The optimization method goal is to find the cutting parameter that increases the machined surface quality while keeping the constraints in mind.

The specific objectives are as follow:

- To develop hybrid particle reinforced aluminum metal matrix composites reinforced with triple nano-particles (SiC, Al_2O_3, and CNT) using stir casting method.
- To investigate physical (notably density, porosity), mechanical (notably tensile strength, flexural strength, hardness, impact resistance) and machinability properties of the prepared composites.
- To optimize the machining parameters (rpm, feed rate) for the prepared composites.

1.4 Scope of this Research

The features of many types of composites were produced and examined by previous scholars. Future manufacturing technology is focused on producing low-cost, light-weight, high-performance items with boundless resources. Single particle reinforced aluminum alloy matrix composites and hybrid particle reinforced aluminum alloy matrix composites were made and investigated in order to improve the performance of aluminum alloy by varying the reinforcements. It is found that a lot of research work has been done on aluminum based metal matrix composite which is reinforced by aluminum oxide and silicon carbide micro-particles, used as single particle reinforcement or bi-reinforcement mixed at various weight ratio. The addition of ceramic particles enhanced the composite's hardness, compressive strength, impact strength, thus the brittle behavior increases, whereas the ductile characteristics decreases. It is not suitable for many type of machine which work under vibrational state. For this adopting carbon nanotubes (CNT) improve the mechanical and tribological performance of the aluminum metal matrix base. It is also seen that nanoparticles reduce the void, formed during the production of composite material. Thus, there is increment in mechanical and tribological properties.

As a result, aluminum metal alloy was used as a matrix material and ceramics particles was used as reinforcements in this study. By integrating those nanoparticle reinforcements in the form of no reinforced, and hybrid particle reinforced at a specified weight composition of each and investigated there properties.

1.5 Organization of the Thesis

The thesis is divided into six chapters including this one and the relevant chapters are organized in the following manner:

Chapter 1 presents the background of the present research. Based on the limitations of the recurrent research demonstrate the objective of the present research while focusing on suggestions from previous experiments. Objectives of the present work are also discussed in this chapter.

Chapter 2 offers the current status of aluminum-based metal matrix composites using various types of reinforcements. This chapter discusses the problems faced by researchers while observing the characteristics of metal matrix composites. And the methods adopted by the researchers for controlling the factors that influence the characteristics. It also presents the general necessities associated with the fabrication process and machining of metal matrix composites. Investigations and findings of previous works are also presented in this chapter. And finally discuss the objectives, scope and limitation of the present research based on the literature.

Chapter 3 deals with the matrix and reinforcement selection and material preparation by the stir-casting mechanism to develop composite materials. The material selection process starts with a statistical method, finding the best suitable option among the all other alternatives available. Then it presents a detailed overview of the fabrication process following the stir-casting method. Chapter 3 also offers a glimpse of the procedure and conditions of material development for testing following international testing standards.

Chapter 4 described the experimental set up, measurement techniques and data processing systems. This chapter deal with the morphological, mechanical, and machining investigations. The experimental findings have been attained by performing the microscopic test, tensile, flexural, impact strength, and hardness, and also the machining test.

Chapter 5 narrates a detailed discussion of the experimental results that have been found through the experimental investigations. In this chapter, also illustrates the development of statistical models for the objective function as well as for the constraint equations in terms of the machining parameters such as cutting speed, feed rate and depth of cut in the drilling and turning process. This chapter also contains a discussion regarding the relation between the chip morphology with the machining parameters.

Chapter 6 contains the concluding remarks with some recommendations for future works. Lastly references are included and appendices are given at the end.

CHAPTER 2
LITERATURE REVIEW

2.1 Introduction

A composite is a structural substance made up of two or more mixed elements that are not soluble in each other and are combined at a macroscopic level. Anisotropic and inhomogeneous materials, composite materials are. Composite materials typically have unique qualities, such as a high strength-to-weight ratio (Chakrapani and Suryakumari 2020, Kumar et al. 2020). Composite materials have two phases: a continuous phase known as a 'matrix' and a discontinuous phase known as 'reinforcements', shown in fig. 2.1. The matrix phase of composite materials is responsible for connecting fibers/reinforcements together, transferring loads and stresses within the composite structure, supporting the overall structure, and protecting the composite from external agents such as humidity, chemicals, and other contaminants (Kumar et al. 2020). Fibers/reinforcements, on the other hand, give composites rigidity, strength, thermal stability, and other structural qualities. The discontinuous phase is often more difficult and powerful than the continuous phase. Shear, low density, and toughness are all provided by the matrix phase (Yang et al. 2021, Prakash et al. 2019).

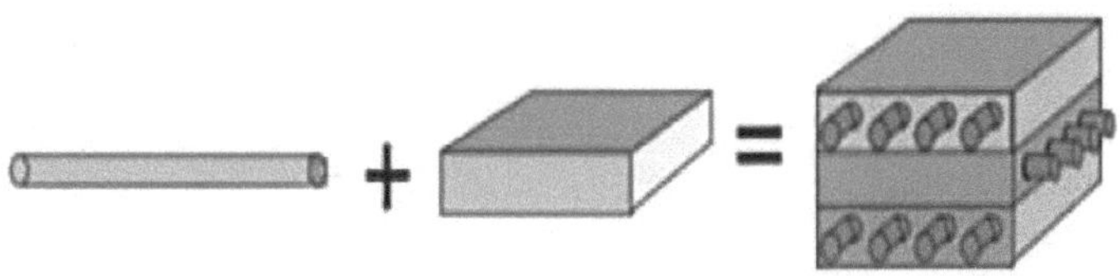

Fig. 2.1: Composition of composite material (Callister 2019).

2.2 Classification and Characterization of Composite Material

The development of composite materials has expanded material property combinations and ranges, and it continues to do so. In general, any multiphase material exhibits a considerable fraction of the properties of both component phases, resulting in a better combination of properties, and is termed as a composite. Better property combinations are fashioned by the appropriate combining of two or more separate elements, according to this principle of combined action. Many composites have property trade-offs as well. Composite materials can be divided into three types based on the nature of their properties, as shown in fig. 2.2. Particle-reinforced, fiber-reinforced, and structural composites are the three primary divisions, each with at least two subcategories.

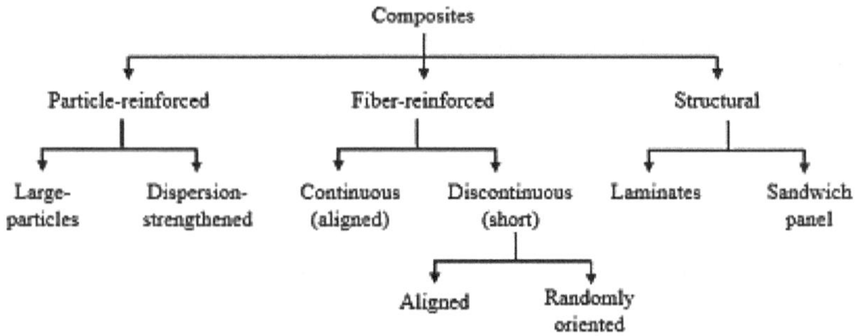

Fig. 2.2: A classification of composite material (Garg et al. 2019).

Metal matrix composites (MMCs), polymer matrix composites (PMCs), and ceramic matrix composites are the three types of structural based composites that can be categorized based on the matrix material employed. Based on the reinforcing material structure, fiber-reinforced composite materials can be classified as particulate composites, fibrous composites, or laminate composites as shown in fig. 2.3 (Garg et al. 2019).

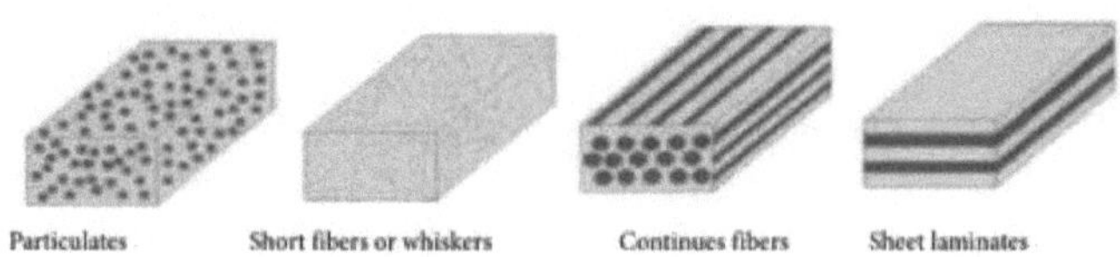

Fig. 2.3: Classification of composite material, a) particulates, b) whiskers/ short fibers, c) fibers, d) layer/plates.

Fig. 2.3(a) shows the matrix reinforced by a dispersed phase in the form of particles in a particulate composite. Because the particles are added at random, they are frequently isotropic. Particulate composites offer benefits such as enhanced strength, high operating temperature, and oxidation resistance, among others. Ceramics and glasses, metal particles, and amorphous materials are all employed as reinforcing particles (Garg et al. 2019). In particle-reinforced composite, the orientation of particles and the size of the particle impact the properties of the composites.

The existence of the particle strengthens the composites, but breaking the particle under a severe load weakens it. The particle's damage process is in charge of this strengthening to-weakening transition, which is determined by the particle's strength, size distribution, and matrix flow behavior. Effective strengthening can be achieved when the particle and matrix are properly matched (Shimizu et al. 2018).

In composites with short fiber reinforcements, as illustrated in fig. 2.3(b), the matrix is reinforced by a dispersed phase in the form of discontinuous fibers (length < 100 x diameter). In composites with long fiber reinforcement, as seen in fig. 2.3(c), the matrix is reinforced by a scattered phase in the form of continuous fibers.

Multilayer composites, as shown in fig. 2.3(d), are fiber-reinforced composites that are made up of many layers with varied fiber orientations.

Composite characteristics are heavily influenced by the properties of their constituent's materials (both the continuous and discontinuous phase), as well as their distribution and interaction. The weight percentage (wt.%) of reinforcement ingredients in the matrix material also affects composite characteristics. Aside from the primary material qualities, the geometry of the reinforcement (form, size, and distribution) has a significant impact on the composite qualities (Shimizu et al. 2018). The orientation of the reinforcement has an impact on the composite characteristics (Ramanathan et al. 2019). The interfacial area is determined by the shape of the reinforcement phase (which can be spherical, cylindrical, or rectangular cross-sectioned prisms or platelets), the size and size distribution (which controls the texture of the material), and the volume fraction, which also plays a role in determining the extent of the reinforcement-matrix interaction. The contribution of a single ingredient to the overall qualities of the composites is determined by the concentration, which is commonly quantified in volume or weight fraction. Composite manufacturing methods and factors also have a significant impact on composite quality (Bhoi et al. 2019).

2.3 Metal Matrix Composites

Metal matrix composites (MMCs) are composites in which metals are used as matrix material. As matrix material; aluminum, magnesium, titanium, and copper alloys and some super-alloys are used. Meanwhile, graphite, silicon carbide, aluminum oxide, boron carbide, molybdenum disulfide, and other common reinforcements are used. Specific stiffness, specific strength, abrasion resistance, creep resistance, thermal conductivity, and dimensional stability, all benefited from the reinforcement (Garg et al. 2019).

Metal matrix composites have a wide range of applications in engineering today. Due to its better qualities than alloys and pure metals. The development of MMCs forced many sectors to embrace metal composites over alloys and pure metals. The followings are some of the industrial applications of metal matrix composites, aluminum alloys are the chosen engineering material for numerous high-performing components used in a variety of applications in the automobile, aircraft, and processing sectors due to their low weight and outstanding thermal conductivity qualities (Pravin and Raj 2019).

Due to their low density, high specific strength, high thermal conductivity, and excellent wear resistance, aluminum-matrix composites are chosen for space-based applications. Thus making them a great alternative for a wide range of technical applications (Kamboj et al. 2012).

The space shuttle's fuselage frame is supported by boron/aluminum tubes. Because of its low thermal conductivity, this composite lowered the weight of the space shuttle by more than 320 lbs. (145kg), as well as the thermal insulation needs (Toor 2017). Carbon-reinforced and silicon-carbide-reinforced aluminum matrix composite is used in building the satellite structure. In recent days, MMCs are also used to make civilian high-performance military aircraft wings, supporting structures, and cargo sections to save weight and decrease fuel consumption (Shimizu et al. 2018).

Metal matrix composites, which are lighter than their metal counterparts, are currently used in vehicle engines. Metal matrix composites are also the preferred material for gas turbine engines due to their great strength and low weight.

MMCs are used by automobile manufacturers in their products. For example, engine components made of an aluminum alloy matrix reinforced with aluminum oxide and carbon fibers have been created, these MMCs are light and resistant to wear and thermal deformation. Sports cars to get high performance, now use composite material for manufacturing external and internal bodies, drive shafts for the transmission, extruded stabilizer bars, and other metal matrix composites are used (Mavhungu et al. 2016).

Aluminum matrix composites are used to make pistons, connecting rods, engine blocks, brake rotors, current collectors, propeller shafts, and brake discs (Grag et al. 2019). Table 2.1 shows the growing applications of aluminum metal matrix composite in the automobile industry.

Table 2.1: Use of Al-matrix composite with different reinforcements in automobile sectors (Mavhungu et al. 2016)

Manufacturer	Composite	Components
Duralcan, Martin Marietta, Lanxide	Al/SiC	Pistons
Duralcan, Lanxide	Al/SiC	Brake rotors, calipers, liners
GKN, Duralcan	Al/SiC	Propeller shaft
Nissan	Al/SiC	Connecting rod
Dow Chemical	Mg/SiC	Sprockets, pulleys, covers
Toyota	Al/Al_2O_3	Piston rings
Hitachi	Cu/Graphite	Current collector
Honda	Al/Al_2O_3-Cf	Engine blocks

Table 2.1 (continued)

Manufacturer	Composite	Components
GM	Al/SiC	Rear brake, driveshaft, engine cradle

Due to its desirable features, the use of metal matrix composite is increasing significantly day by day.

2.4 Recent Research on Aluminum-Based Metal Matrix Composites

In metal matrix composite, a matrix gives the dimensions of the MMCs as the medium of binding and holding immersed materials into a fixed dimension. The matrix phase is mainly single material or alloy of materials, where the reinforcements that are added, are dispersed throughout the matrix. Aluminum metal matrix composites (AMMCs) differ from other metal matrix composites in that, they have a higher strength-to-weight ratio. This section provides a thorough review of previous work on aluminum metal matrix composites. Single particle-reinforced aluminum metal matrix composites and hybrid particle-reinforced aluminum metal matrix composites will be discussed.

2.4.1 Particle-Reinforced Metal Matrix Composites

On the particle's reinforced metal matrix composites, the strengthening effects, and the damaging process, many attempts have been done including experimental and numerical research. In general, from various research, it can be said that composite strength improves as particle size decreases. However, damage occurs continuously during deformation and may reduce the strength of the composites.

To study the effect of particle size on composite properties, an aluminum-based metal matrix composite reinforced with 10 vol.% silicon carbide particle was prepared, where two variations of particle sizes of 40μm and 2μm. The accumulative roll bonding (ARB) procedure is used in the production of this aluminum metal matrix composite reinforced with silicon carbide particles. When comparing, MMCs with 40μm particle sizes to MMCs with 2μm particle sizes, it was discovered that the microstructural evolution in MMCs with 40μm particle sizes was more prominent. Furthermore, the composite strip with a particle size of 40μm became more homogenous, high-bonding quality, and porosity-free sooner than the strip with a particle size of 2μm (Jamaati et al. 2010).

El-kady and Fathy on their work observed the effect of silicon carbide particle size on the aluminum metal matrix composite. The size of silicon carbide was used 70nm, 10μm, 40μm and aluminum powder particle size was 60μm. The powder metallurgy method was followed to produce the composite. The composite's electric and thermal conductivity decreased as the amount of silicon carbide wt.% was increased, but increased as the silicon carbide particle size was raised. Brittle properties like hardness and compressive strength also increased as the wt.% of reinforcement particles was increased. On the basis of size, brittle properties increase as the particle size decreases (El-Kady & A, Fathy, 2013)

Song et al. worked with aluminum oxide-reinforced aluminum metal matrix composite. They use 20 vol.% of aluminum oxide to manufacture the composite and observed the reinforcement shape effects (spherical and angular) on the prepared composite fracture behavior and ductility. Their experiment shows, the composite with spherical-shaped reinforced particles, has a lower yield strength but higher ductility and a slower work hardening rate than the angular particle reinforced composite (Song et al., 1996).

Williams et al. using X-ray synchrotron tomography, investigated the damage behavior of silicon carbide particle reinforced 2080-Al alloy matrix composite. And they concluded that the fracture is more likely in particles with greater size and aspect ratio. The particle's strength appears to be a stronger function of the aspect ratio. The composite damage zone was restricted to a very small volume near the fracture plane. Both void growth and particle fracture occurred within a 1mm radius of the fracture plane (Williams et al., 2010).

The effects of reinforcing particle shape and interface strength on the deformation and fracture behavior of an aluminum metal matrix composite reinforced by aluminum oxide particles are investigated numerically by Romanova et al. They varied the reinforcing particle shaped from spherical to irregular shape. A dynamic stress-strain state is achieved in the vicinity of the matrix/particle interface due to structural heterogeneity, and areas undergoing tensile deformation are generated near the interface in both tension and compression. The macroscopic strain of fracture initiation is largely influenced by the geometry of the reinforcing particles. The greater the roughness of the interaction, the sooner the fracture begins and two fracture mechanics they observed: interface de-bonding and particle cracking (Romanova, Balokhonov and Schmauder, 2009).

Slipenyuk et al. in their study of aluminum metal matrix composite reinforced by silicon carbide and manufactured by powder metallurgy investigated the effect of reinforcement particle size with two variations 3μm and 14μm with volume fraction variation of 0 to 20 vol.% on the microstructure and mechanical properties (yield stress, tensile strength, elongation to fracture, and young modulus). They conclude that, on the condition of a uniform reinforcement distribution, decreasing the reinforcing particle size provides for higher yield stress and tensile strength as well as greater material fabric ability. For lower reinforcement concentrations, a homogenous distribution of reinforcement particles is achievable, and determining these concentrations becomes a crucial stage in the material decomposition process (Slipenyuk et al., 2006).

Metal matrix composites are referred to as single reinforced metal matrix composites when only one reinforcement material is used. As stated below, many researchers use aluminum alloy as a matrix material with single reinforcement to create single reinforced composites.

Votarikari et al. investigated the mechanical and material properties of Al-6061 metal matrix composites reinforced by 300-350nm sized Nano-silica particles. The composite was made adopting powder metallurgy technique by changing the reinforcement from 1 to 15 wt.% in 2 wt.% increments. After that, the produced the composites, microstructural, hardness, and tensile characteristics were determined with and without the inclusion of Nano-silica particles. The composites microstructural analysis revealed uniform grain distribution and modest grain refinement in the specimen. In addition, when compared to standard unreinforced 6061-Al matrix, it was found that, the yield strength of the Al-Nano silica composite increased up to 10 wt.% Nano-silica component, by weakening bonds between Al-6061 alloy matrix and Nano-silica powders with greater silica content more than 10wt%, the same silica particles turn the ductile Al-6061 matrix into a brittle matrix. And they suggested to limited the addition of Nano-silica particles to 10wt% for maximum efficiency and density also increases as the reinforcement wt.% increased (Votarikari, no date).

Kuwarmausam et al. investigated aluminum metal composites reinforced with boron carbide, that were manufactured using the stir casting method. AMMCs was made with reinforcing boron carbide at weight percentage of 5%, 10%, 15% and 20%. Optical micrographs revealed a uniform dispersion of boron carbide particles in the AMMCs. The tensile strength and hardness of the composite have improved as the weight percentage of boron carbide particles has grown. At the 20 wt.% of boron carbide, the maximum ultimate tensile strength and hardness are 187.363 MPa and 185 HV, respectively. Their study indicated that, as the weight percentage increases the brittle property of the composite (Kuwarmausam et. al, 2021).

A hybrid metal matrix composite is created when two or more reinforcements are used in the fabrication metal matrix composite (Rouhi, Moazami-goudarzi and Ardestani, 2019). Due to differences in matrix and reinforcement qualities, the created material characteristics were altered. Metal matrix hybrid composites are second-generation composite materials that have the potential to replace single-reinforced composites by increasing characteristics and meeting the needs of sophisticated engineering application (Madhukumar and Umashankar, 2018, Rouhi et al. 2019). Various researchers perform research on hybrid metal matrix composites and hybrid reinforced aluminum alloy matrix composites, which are discussed as follow.

The microstructure, hardness, and corrosion characteristics of aluminum based composites reinforced with silicon carbide (SiC) and Ferro-titanium (TiFe) particles were studied by Akinwamide et al. In the potentio-dynamic polarization and potentio-static test of their study, the specimens reinforced with particles of 5wt% SiC + 2wt%

TiFe and 5wt% SiC + 5wt% TiFe showed the greatest improvement in corrosion resistant. This showed, the inclusion of these reinforcements also enhanced the pace at which oxide layers developed on the composites surface, increasing corrosion resistance (Akinwamide, Akinribide and Olubambi, 2020).

By using stir casting and compo-casting process, David et al. synthesized and analyzed Al-6061 metal matrix composite reinforced with fly-ash and silicon carbide particles. Different weight percentages of silicon particles and a fixed weight percentage of fly-ash were used to strengthen the AMMCs. The constructed AMMCs microstructure and mechanical properties were investigated. With an increase in the wt.% of silicon carbide particles in the aluminum matrix and a constant wt.% of fly-ash in the aluminum matrix, mechanical characteristics such as hardness and tensile strength were improved. The ultimate tensile strength and maximum hardness are 80 VHN and 213 MPa, respectively (Selvam and Dinaharan, 2013).

Suresh et al. looked into the effect of graphite on the mechanical behavior of an Al-6061-titanium bromide hybrid composite. Stir casting was used to make Al-6061 reinforced hybrid composites with 2wt% graphite particle size of 20-25μm and 0wt%, 10wt% and 20wt% titanium bromide particle size of 1-5μm. Following that, mechanical behaviors such as hardness, tensile strength, and fatigue behaviors were examined utilizing acoustics emission. The composite with the composition Al-6061-20% titanium bromide and 2wt% graphite has the best mechanical behavior. At a 20wt% titanium bromide with 2wt% graphite composition, the maximal value of hardness of is 91.4 HV (Suresh et al., 2014).

Partheeban et al. used powder metallurgy to investigate the mechanical behavior of Al-6061 alloy, Al-6061 and 10wt% titanium bromide, and Al-6061 with 10wt% titanium bromide and 1wt% graphite, and Al-6061 with 10wt% titanium bromide and 2wt% graphite hybrid composites. The particle sizes of the aluminum and titanium bromide are 30-50μm and 1-10μm, respectively. And the graphite particle size is 25-50nm. The addition of titanium bromide and Nano-graphite reinforcements to the Al-6061 matrix enhanced the hardness. With the addition of titanium bromide and Nano-graphite, the tensile strength of the material improved as well, Al-6061 metal matrix reinforced with 10wt% titanium bromide and 2wt% graphite, composites has the greatest hardness value of 79.04 HV and the greatest ultimate compressive strength value of 322 MPa (Anand Partheeban et al., 2015).

Johny et al. developed and investigated the properties of a stir cast hybrid aluminum metal matrix composite. Aluminum 6061 was chosen as the metal matrix, while zirconium dioxide and aluminum oxide particles with an average size of 55-65μm were used as reinforcements. The composite is made up of 90wt% aluminum 6061, 5wt% zirconium dioxide and 5wt% aluminum oxide, all of which are confined to a 10wt% proportion. The ASTM-E8 standard was followed for cutting the tensile strength test specimens, and the results were tabulated. The tensile value has improved significantly, with the greatest UTM value of 227.3 MPa. Vickers micro hardness tester was used for the hardness test, and the maximum value was 82.9 HB at 200 kg force (Akinwamide, Akinribide and Olubambi, 2020).

Viney et al. investigated the mechanical properties of Al-6061+ Mg 4wt%+ Fly-ash and Al-6061+ Mg 4wt%+ Graphite 4wt% + fly-ash hybrid metal matrix composite manufactured using the stir casting process, as well as the effect of sliding velocity on wear parameters. The composition of reinforcement wear was as follows. In the first case, Al-6061 was chosen as the base metal with 4wt% Mg and varying compositions of fly ash (10wt%, 15wt%, and 20wt%) as reinforcements. In the second case, Al-6061 was chosen as the base metal with 4wt% Mg and 4wt% graphite as the fix and varying composition of fly ash (10wt%, 15wt%, and 20wt%) as reinforcements. The addition of fly ash increased the tensile strength of the material. When graphite was added, the tensile and hardness of the material decreased. The composite with 4wt% Mg and 15wt% fly ash had the highest tensile strength of 124 MPa, whereas, the composite with 4wt% Mg and 20wt% fly ash had the highest hardness with a value of 71.3 HV (Kumar, Gupta and Batra, 2014).

Halil et al. looked into the wear and mechanical properties of Al-6061 metal matrix alloy (particle size is less than 100μm) + silicon carbide (particle size is less than 8μm) + boron carbide (particle size is less than 10μm) hybrid composite made by powder metallurgy. It explains a new method for improving the mechanical properties of hybrid composites (Al-SiC-B_4C) made via powder extrusion. T6 thermal treatment was applied to the extruded samples. The terms of density, microstructure, hardness, transverse matrix transverse rupture strength, tensile strength, and wear resistance are investigated once the composite materials are formed. The hardest composites with a 12wt% B4C reinforcement had a hardness of 732HB. The composite material with 12wt% SiC particle reinforcement had the maximum tensile strength of 405MPa (Halil et al., 2019).

The physical and mechanical properties of Al6061+Silicon Carbide (SiC) (mesh size 220) +Coconut shell ash (CSA) (40–80 μm size) hybrid composites manufactured by stir casting were examined by Satheesh and Pugazhvadivu. The Al6061-SiC composite was made by mixing 10% SiC into the Al6061 melt. The density of the Al6061-SiC-10wt% CSA hybrid composite is 6.58%, which is lower than the density of the Al6061-SiC composite. The hardness and tensile strength of the Al6061- SiC-8wt% CSA hybrid composite rose by 46.31% and 47.31%, respectively. The maximum hardness and tensile strength of Al6061+SiC-8wt% +CSA hybrid composite improved to 1002VHN and 200MPa, respectively (Satheesh and Pugazhvadivu, 2019).

Madeva et al. used liquid metallurgy to synthesize and analyze microstructural and mechanical properties of Al6061, Al6061+ 9wt% SiC particles of size 90-125μm, and Al6061+9wt% graphite particles of size 90-125μm composites. Both SiC and Graphite composites were tested for density, hardness, ultimate tensile strength, yield strength, and percentage elongation. The particle distribution in these composites was uniform, according to the microstructural examination. The addition of 9 wt.% SiC to Al6061 enhanced the density and hardness of the material. In comparison to the Al6061 matrix system, Al6061+9 wt.% SiC composites showed a 25.32% increase in ultimate tensile strength. At 9 wt.% SiC, the maximum value of hardness is 1502VHN, and at 9 wt.% graphite, the highest value of ultimate tensile strength is 1932MPa (Nagaral et al., 2018).

Subramani et al. investigated the properties of a stir cast A6061/ (Glass Fiber + AL_2O_3 + SiC + B_4C) reinforced hybrid composite. The reinforcing particles are 25μm in size on average. A6061 and its reinforcing particles, such as aluminum oxide/glass fibers/SiC/B_4C (4wt% and 5wt%), were created with and without glass fiber composite. Microstructural and mechanical testing, such as micro hardness, wear, and tensile testing, were then performed. The greatest ultimate tensile strength (125MPa) and hardness (49.7BHN) were measured in samples with glass fibers + 5% other reinforcements and samples without glass fibers + 5% other reinforcements, respectively (Nanjan and Janakiram, 2019)

Sagar and suresh investigated the effect of beryl reinforcement on the mechanical and tribological properties of Al-2024 matrix material. Cold extrusion process was used to construct the composite specimens reinforced with 2wt.% to 12wt.% in 2wt.% increments. They compare the result with standard beryl reinforced aluminum metal matrix composite processed by casting, and found the ultimate tensile strength and hardness number for the composites prepared by cold extrusion is 1.95 to 2.35 times higher than the composites prepared in casting and also the cold extruded composite with 6wt% beryl reinforcement shows the best result among others. They conclude, in cold process the particle distribution was even and thus the grain structure was better than the casting process (Sagar K G and Suresh P M, 2020).

Siddesha et al. studied the effect of titanium oxide reinforcement in aluminum 2024 metal matrix composite based material. They varied the weight percentage of titanium oxide by 2%, 4%, 6% and 8%. Stir casting technique was used to construct the composite. Stir casting process helps the reinforcement particles distribution properly and from experiment they found the tensile strength, hardness and impact strength of the material increases as the weight percentage of reinforcement particle increased (Siddesha, 2016).

Bhat and Ganesh used stir casting process to develop Al-6061 metal matrix composite reinforced with silicon carbide particle. They use 5wt% of silicon carbide, size of 50μm. The main objective is to investigated the wear behavior on the composite due to inclusion of reinforcement particle. In their experiment, to observe the wear resistance, they varied the load from 5N to 200N and RPM varying from 200 to 1500. As a result, they concluded that a reinforced silicon carbide metal matrix composite outperforms an unreinforced Al-6061 alloy. The wear resistance increases as the weight percent of silicon carbide reinforcement increases (Bhat and Kakandikar, 2019).

Pitchayyapillai et al. produced and characterized the property of Nano-silver reinforced Al-6061 metal matrix composites manufactured by stir casting process. The Nano-silver reinforcement is used with various percentage of 1wt% and 2wt%. The size of the Nano-silver varies from 60 to 100 nm. In the fabrication process via stir casting method, the matrix material molten at 760-degree centigrade temperature and after mixing the reinforcement it stirred for 20 min at 550 r/min. Microstructure, micro-hardness, ultimate tensile strength, and compressive strength were measured after manufacture. According to their, the hardness of manufactured composite was increased with the increasing of reinforcement weight percentage compared to standard specimen. The density of standard specimen was 2.680 gm/L, whereas at 1wt% Nano-silver reinforced composite density was 2.705 gm/L and 2wt% Nano-silver reinforced composite density is 2.740 gm/L. So, the density improved proportionally as the weight percentage of reinforcement increased. Due to the increment of reinforcing particle, the ductility of the aluminum alloys also increased significantly (Pitchayyapillai et al., 2017).

Saheb et al. used carbon nanotubes as the reinforcement particles with Al-6061 and Al-2124 metal matrix composite, to evaluate the effect by including the Nano-particles. Wet ball milling spark plasma technique was used to fabricate the composite material. Discovered by Iijima, carbon nanotubes gain its popularity due to exceptionally strong characteristics (Sumio Iijima, 1991). To get the most out of this highly powerful reinforcement, dispersion of CNTs demands careful planning of dispersion strategies and routes. The use of combination of sonication and wet milling in their study increased CNT dispersion, resulting in uniform distribution of CNTs in a composite of containing 1wt% of CNTs. However, significant accumulation was detected in the composite containing 2wt% CNTs, resulting in low composite hardness. Their study suggested, CNTs concentration of 1wt% is optimal for producing exceptionally dense and hard Al-6061 and Al-2124 nanocomposites (Saheb et al., no date).

Venkatesan and Xavior studied on 7xxx series aluminum alloy reinforced with graphene. They used AA-7050 metal as the base matrix material and reinforced it with graphene particle, which manufactured via stir and squeeze cast procedure. The experiment took into account certain parameters such as melting temperature (775, 800 and 825 degrees centigrade), stirring speed (300, 400 and 500 RPM) and graphene content (0.3, 0.5 and 0.7 wt.%) with three levels for better comparison. The amount of influence of the parameters on the specimen's tensile strength was determined using analysis of variance (ANOVA). The size of the graphene particle varied from 50-100nm. Results shows, as the stirring speed increased, the material composition was better. The stirring process avoids the floating of reinforcement particles over the molten matrix metal and ensure proper distribution, this results in greater particle bonding in the matrix. The result on their study showed the tensile strength was most influenced by inclusion of graphene particle, as graphene wt.% increased the tensile strength also increased (Venkatesan and Anthony Xavior, 2019).

Singh et al. studied the effect of reinforcing boron carbide on AA-6061 metal matrix base. The boron carbide ceramic serves as a strong reinforcement in the base matrix, improving the composites mechanical properties. The composites were created using the stir casting process, and boron carbide with a particle size of 50μm was reinforced in the AA-6061 by changing weight percentages from 1% to 15% in 5% increment. The final results showed, aluminum metal matrix composite with 1wt% boron carbide reinforcement had Rockwell hardness number 62.6 HRB, with 5% boron carbide reinforcement had Rockwell hardness number 66.3 HRB, and 15% boron carbide reinforcement had had Rockwell hardness number 78.5 HRB. Thus the impact strength also increased as the boron carbide weight percentage increased (Singh et al., 2021).

Hashim et al. studied the benefits of heat treatment on improving the mechanical characteristics of an aluminum-silicon alloy metal matrix base reinforced by multiwall carbon nanotubes (MWCNT). Stir casting method was utilized to fabricate the composite, during mechanical stirring, the vortex condition in the liquid matrix managed to break and prevent the production of helical arms while also assisting the distribution of MWCNT throughout the matrix. Due to load transfer and a high density of dislocation forms around themselves, MWCNT reinforcement aids to restrict dislocation mobility in the matrix. By reducing micro porosities in the matrix through non-dendritic matrix movement during compaction, the thermoforming method has boosted the strength even more (Hashim et al., 2021).

Youssef and Sayed investigated the effect of several processing parameters on the tensile performance of Al-10Sb/SiC composites, such as particle size and reinforcement percentage. Via stir casting technique, they mix the reinforcements with the metal matrix at three particle sizes of 115, 225, and 350 μm applied, with three reinforcing weight fractions of 3%, 5%, and 9%. To stir and mix the molten composite, the impeller was circulated at a speed of 350-400 RPM for five minutes. The yield strength of the 115 μm size particles was increased by roughly 21% when 9wt% silicon carbide particles were added. Similarly, for the same particles weight concentration the yield strength increased by 18% and 13% for the 225 μm and 350 μm particle size, respectively. On the other hand, they found a significant decrease in ductility, which was more obvious with the addition of fine particles and a high mass fraction (Youssef and El-Sayed, 2016).

Bhasker et al. used stir casting method to manufacture and characterize the alumina reinforced Al-6061 metal matrix composite. The aluminum alloy 6061 was reinforced with different weight percentages of aluminum oxide particles (5%, 10%, 15% and 20%) in their study. Microstructures, micro hardness, and ultimate tensile strength of AMMC were measured after manufacture. According to their findings, aluminum oxide particles were dispersed uniformly throughout the aluminum matrix. It was also discovered that when the percentage of reinforcement grew from 5wt% to 20wt%, tensile strength, ultimate tensile strength, and hardness all improved. The proportion of elongation, on the other hand, was reduced. With the addition of 20wt% aluminum oxide, the greatest ultimate tensile strength and hardness were recorded, with a maximum value of 310 MPa and 89.91 VHN respectively (Kandpal, Kumar and Singh, 2017).

Dhanashekar et al. studied the mechanical characteristics and wear behavior of powder metallurgy fabricated AA-6061 metal matrix silicon carbide reinforced composites. The varied silicon carbide reinforcement in 2.5wt%, 5wt% and 7.5wt% weight percentage. Results showed, the compressive strength and micro Vickers hardness was increased as the reinforcement weight percentage increased. By microscopic study, they found the porosity in the fabricated composite specimen was increased as reinforcement percentage increased. Which showed the limitation of powder metallurgy process to fabricate metal matrix composites (Dhanashekar , P. Loganathan, S. Ayyanar, S.R. Mohan, T. Sathish, 2020).

Sivananthan et al. fabricated metal matrix composites with silicon carbide particle reinforcement made from Al-6061 aluminum alloy utilizing the stir casting method. The weight percentage of silicon carbide particles ranged from 0 to 4% of 44μm in size. The stirring process is done at 350 centigrade temperatures and mixing blade speed was 650 RPM for better mixing. The finding showed that, when the weight percentage of silicon carbide particles in the aluminum 6061 alloy grew, the hardness, tensile strength, and compression strength of the samples increased. When compared to aluminum 6061 standard alloy specimen, hardness, tensile strength, and compression strength rose by 25%, 25.6% and 12%, respectively for 4wt% silicon carbide reinforced composite (Sivananthan, Ravi and Samuel, 2020).

Maurya et al. experimentally investigated the effect of silicon carbide reinforcement concentration on the aluminum alloy Al-6061 metal matrix base material. By using the stir casting method, composites with different silicon carbide (0wt%, 1wt%, 2wt%, 3wt% and 5wt%) were created by them. The density of the Al-6061 metal matrix silicon carbide reinforced composite was increased by 1.4% when 5wt% silicon carbide was added for reinforcement. Similarly, for 5wt% of silicon carbide reinforcement added to the manufactured composite, the hardness increased by 12.5%. The Al-6061 metal matrix and 5wt% silicon carbide reinforced composite's tensile strength was increased to 288MPa. The tensile strength of a produced composite had been increased due to the presence of hard silicon carbide particles (Maurya et al., 2019).

Idrisi et al. statistically analyze the wear characteristic of Al-5083 metal matrix composites reinforced by silicon carbide Nano-particle. Stir casting was used to create aluminum matrix composites with varied percentages of silicon carbide nanoparticles (0wt%, 1wt% and 2wt%). Taguchi's approach was used to optimize the wear parameters. For Al-5083 composite reinforced with silicon carbide nano-particles, it was discovered that the experiment time has the greatest impact on wear rate, followed by silicon carbide wt.%, and finally the applied load (Idrisi et al., 2018).

Sahu et al. manufactured particulate Nano-composite material, using an improved ultrasonic assisted stir casting technology to integrate Nano-sized aluminum oxide particles in aluminum alloy 6061 metal matrix base. The amount of aluminum oxide Nano-particles in the composite was changed from 0 to 3wt% weight, and the influence of various parameters on wear rate was investigated by them. Result showed, heat treated alloys and composites had improved wear characteristics than non-heat treated alloys and composites. The temperature of a composite specimen varies in direct proportion to the applied load, and with increasing stress, the coefficient of friction of composites drops (Sahu et al., 2015).

Idrisi and Mourad performed a comparison between conventional stir casting and ultrasonic aided stir casting, where both these techniques were used to create aluminum metal matrix composites (AMMCs). Different concentrations of silicon carbide (3wt%, 5wt%, 8wt% and 10wt%) micro particles with a size of 40μm were used to make AMMCs. The results showed, the ultrasonic aided stir casting method has been demonstrated to be superior to the traditional stir casting method. This is owing to adequate silicon carbide mixing in the aluminum alloy and the absence of voids (Idrisi and Mourad, 2019).

Canakci et el. on their study made Al-composites using mechanical alloying technique and traditional powder metallurgy methods with aluminum oxide particles up to 15wt% as reinforcement, and investigated microstructural changes in aluminum oxide particles, and investigated the impact of mechanical alloying processing on the distribution of aluminum oxide particle reinforcement and its properties. Microscopic study showed, by mechanical alloying process, the reinforcement dispersion uniformly. It was discovered that, increasing the volume fraction (reinforcement ratio) results in higher hardness (Canakci, Varol and Ertok, 2012).

Kumar et al. investigated nanocomposites made of Al-6061 and graphene. Graphene, can be used as a Nano-filler and perfect reinforcement for traditional structural materials, according to research. In the aluminum matrix, graphene with varied filling contents (0.25wt%, 0.5wt%, 0.75wt% and 1.0wt%) is used as reinforcement. At its optimal graphene concentration in the produced composite, the investigation demonstrates that adding graphene increases hardness values while decreasing diametrical expansion. At 0.5wt%, the greatest micro hardness values are showed. The minimum expansion was recorded at 1wt% graphene at a quenching temperature of 450 degrees centigrade, while graphene sintered at 500 degrees centigrade has a value of 84 HV (Prashantha Kumar and Anthony Xavior, 2017).

Bragaglia et al., in 3.5% sodium chloride, studied the corrosion resistance properties of two distinct metal matrix composites based on Al-6061 and Al-2618 aluminum alloys reinforced by 20wt% aluminum oxide particles by stir casting technique and compared the result. Results showed, the inclusion of aluminum oxide reinforcement, enhance the kinetics of precipitation of the composites in the chemical solution of sodium chloride (Bragaglia, Montanari and Montesperelli, 2019).

Senapati et al. experimentally investigated the mechanical characteristics of aluminum-silicon alloy reinforced with 10wt% fly-ash at different stirring speeds. The stirring speed was changed between 600 and 650 RPM while all other parameters remained constant. Reports showed, as the stirring speed increased, the impact strength of the fabricated composite is higher compared to the composite, fabricated at lower stirring speed. It may be the reason, as the stirring speed increased, the particles were properly distributed. The results of the experiments show that, the stirring speed during the liquid casting process is significant in the manufacturing of metal matrix composites, especially when the material's mechanical properties need to be improved (Senapati et al., 2020). And use response surface methodology to optimize the proper machining parameters (Senapathi, Raju and Rao, 2017).

Gowri et al. evaluated the effect of artificial aging on stir cast Al-6061-boron carbide and Al-6061-silican carbide composites in order to improve hardness. By altering the weight percentage of reinforcements (2wt%, 4wt% and 6wt%) on Al-6061 alloy, this study studied the individual effects of boron carbide and silicon carbide on the improvement in hardness with an average size of 35-40μm. The influence of precipitation hardening on mechanical characteristics at 100, 150, and 200 degrees Celsius was also examined. With an aging duration of 9.30 hrs. at 100 degrees Celsius, a peak hardness of 85 BHN is recorded. Peak hardness increases to a maximum value with decreasing aging temperature and increasing aging duration, according to the findings. When compared to 50 BHN of Al-6061 alloy, the cast hardness of Al-6061 with 6wt% boron carbide composite is 74 BHN and Al-6061-6wt% silicon carbide composite is 69 BHN (Gowri Shankar et al., 2018).

The mechanical property of a friction stir treated Al-6061-boron nitride surface composite were studied by Akshay et al. The percentage of boron nitride was changed in steps of 3wt% from 0 to 6wt%. Microstructure as a function of the number of passes and the reinforcing content (optical and SEM), Al-6061 was tested for hardness and tensile behavior. Increases in boron nitride concentration and number of passes result in a considerable rise in hardness. When compared to friction stir processed Al-6061 alloy, Al-6061-3wt% boron nitride increased by 14% and Al-6061-6wt% boron nitride increased by 20% (Akshay et al., 2018).

Prabhu et al. studied the microstructure and mechanical properties of rutile-reinforced AA-6061 matrix composites made by stir casting. The composite was made by altering the mass fractions of rutile average size of 10μm particles in the AA-6061 matrix to 1wt%, 2wt%, 3wt% and 4wt%. When compared to the base material, rutile reinforced AMMCs had greater tensile strength and hardness. By increasing the mass fraction to rutile particles in the composites, the characteristics of the composites improved. The tensile strength of the material dropped after 3wt% rutile particles were added. The hardness of the specimen rises by up to 50% when rutile particles are added to it compared to the base matrix alloy (Prabhu et al., 2019).

Kishore et al. used the Taguchi approach to investigate surface roughness in turning of an in-situ Al-6061 metal matrix reinforced by titanium chloride composite. A stir casting system is used to produce Al-6061-10wt% titanium chloride MMCs rod. Cutting speed, with a percentage contribution of 43%, was shown to have the greatest impact on response surface roughness, followed by depth of cut and feed rate, with percentage contribution of 31% and 26%, respectively (Kishore et al., 2018).

Ashok et al. investigated the tensile and compression behavior of Al-6061 boron carbide reinforced MMCs that were cast using the conventional stir casting method. At a processing temperature of 750 degrees Celsius, the composite is made up of 9wt% boron carbide particles with a mean size of 8μm. Tensile, compressive and hardness microstructural tests were done after the composite was manufactured. Microstructural examinations revealed a rather homogeneous dispersion of boron carbide particles in the Al-6061 matrix. The Al-6061-9wt% boron particles composite's hardness has been enhanced by 34.9%. After adding 9wt% boron carbide particles to the Al-6061 alloy, the yield stress and ultimate tensile strength improved by 15.47% and 9.5%, respectively. And the overall strength improved by 24.25%. The ductility of the composites, on the other hand, is reduced when boron carbide particles are added. As a result, the Al-6061-boron carbide system might be employed in structural applications to substitute steel and other ferrous materials (Kumar et al., 2018).

The mechanical and wear properties of Al-6061 red mud composites were examined by Gangadharappa et al. Stir casting was utilized to make Al-6061 red mud composites specimens with 5wt%, 10wt%, 15wt% and 20wt% of red mud 30-40nm size particles as reinforcement. After the composite was created, tests on hardness, wear rate, and friction coefficient were carried out. When the red mud content in the base alloy was increased, the hardness raises under all heat treatment conditions. With increased red mud concentration, the peak age timing that delivers the maximum hardness was shown to decreases (Gangadharappa et al., 2018).

Madhukumar and Umashankar investigated glass particle reinforced aluminum alloy Al-6061 metal matrix composites made by melt stirring. The composites were made with varied reinforcing volume fractions of 250 micros (3wt%, 6wt%, 9wt% and 12wt%). Mechanical tests such as hardness and tensile strength were performed on the prepared composite specimens using a Vickers hardness and uniaxial tensile testing device. As a result of the findings, the hardness and tensile strength of the material rises with an increase in wt.% of reinforcement up to 9wt% and decreases for 12wt% (Madhukumar and Umashankar, 2018).

2.4.2 Various Fabrication Processes of Metal Matrix Composites

A matrix gives the dimensions of the MMCs as the medium of binding and holding immersed materials into a fixed dimension. The matrix phase is mainly single material or alloy of materials, where the reinforcements that are added, are dispersed throughout the matrix. The commonly used matrix and reinforcements in the manufacturing of metal matrix composites are shown in the figure 2.4.

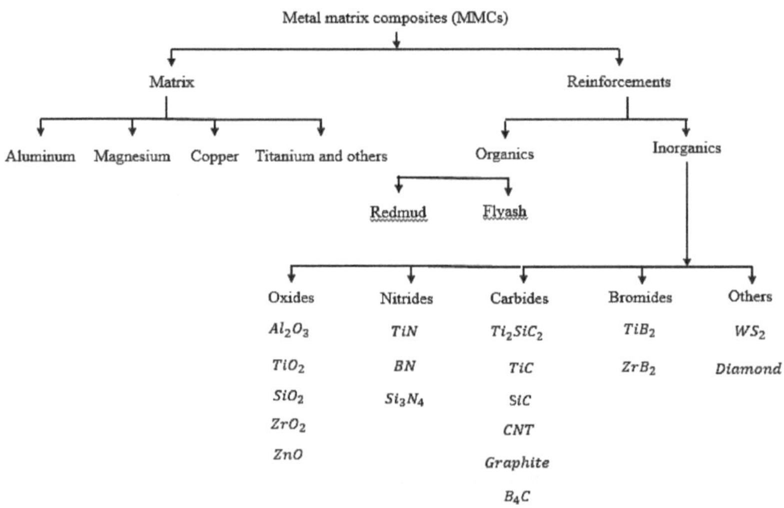

Fig. 2.4: Commonly used matrix and reinforcement materials in MMCs production (Ramanathan, Krisnan and Muraliraja, 2019).

Primary procedures such as processing the metal matrix in a liquid or solid form, as well as others (including semi-solid, in situ and others), can be used to categorize metal matrix composite production methods. Metal matrix composites can be classified as solid state processing, such as powder metallurgy, or liquid state processing, such as liquid metallurgy. Stir casting, stir squeezed casting, and other liquid state processing techniques (Ramanathan, Krishnan and Muraliraja, 2019). The advancement in the fabrication methods along with the provision of adding a wide range of reinforcing materials, enable the MMCs for mass-production with numerous application. Stir casting is the most productive, conventional, simple, economical and straightforward of the numerous types of MMCs production processes (Kareem et al., 2021).

Major challenges while processing the metal matrix composites are as follow:

- Non-homogeneous distribution of the reinforcing elements.
- Deficient matrix material dispersion of the reinforcing elements.
- Matrix and reinforcement have a weak interfacial connection.
- The problem of wettability between the reinforcement and the matrix material.
- The composite material's chemical, thermal and mechanical stability.

Table 2.2 compares several production processes in terms of various characteristics.

Table 2.2: Use of Al-matrix composite with different reinforcements in automobile sectors (Mavhungu et al. 2016)

Process	Method		Range of shape and size	Damage of Reinforcement	Cost
Solid state processing	Powder metallurgy and sintering	Conventional sintering	Wide range but restricted side	Reinforcement fracture	Expensive

Table 2.2: Continued

Process	Method		Range of shape and size	Damage of Reinforcement	Cost
		Microwave assisted sintering	Limited shape	-	Expensive
		Spark plasma sintering	Limited shape	-	Expensive
		Hot compaction	Limited shape	-	Expensive
	Laser deposition process		Wide range but restricted size	-	Expensive
	Deformation processing		Limited shape but large size	-	Expensive
	Powder thixo-forming		Limited shape	-	Expensive
Liquid state processing	Casting	Stir casting	Wide range of shape and large size	No damage	Least expensive
	Centrifugal casting		Limited shape but large size	No damage	Moderate

Table 2.2: Continued

Process	Method	Range of shape and size	Damage of Reinforcement	Cost
	Gravity die casting	Limited shape but large size	No damage	Least expens ive
	Spray casting	Limited shape but large size	Severe damage	Expen sive
	Squeeze stir casting	Wide range but restricted shape	No damage	Moder ate
	Melt infiltration	Limited shape	Severe damage	Expen sive
	Microwave assisted hot pressing	Limited shape and restricted size	Reinforcement fracture	Expen sive
Newly developed methods	Sandwich processing	Wide range but restricted size	Severe damage	Expen sive
	Disintegrate melt deposition	Limited shape	Severe damage	Expen sive

Table 2.2: Continued

Process	Method	Range of shape and size	Damage of Reinforcement	Cost
	Friction stir processing	Limited shape	Reinforcement fracture	Expen sive
	Vapor deposition process	Limited shape	-	expens ive

Stir or squeezed-stir casting process is widely utilized to manufacture high-quality single or hybrid reinforced metal matrix composite materials. In stirring casting process, the melting temperature of the matrix, die temperature, string speed, string time, reinforcement wt.%, reinforcement preheating temperature, and other process factors all play a significant influence in the fabrication of high-quality metal matrix composite (MMCs).

One feature that has a considerable impact on the properties of metal matrix composites is wettability. The capacity of a liquid (molten matrix) to spread over a solid surface (reinforcement surface) is known as wettability. It also refers to the degree to which a liquid and a solid are in close proximity. To improve the wettability of the reinforcement particles with a liquid matrix material, a variety of approaches have been used (Kamboj, Kumar and Singh, 2013).

- Wettable metals like nickel and copper are used to coat the reinforcing metals
- The addition of reactive alloying elements such as Mg, Ca, Ti, NaC_3, or Zr to the molten matrix alloy
- Wettability is also improved by preheating the reinforcement particles before adding it to the molten matrix.
- The use of a mechanical string results in a high level of wettability.

- Irradiation of the melt with ultrasonic waves.

A mechanical stirrer is necessary in the stir casting setup to create a vortex for the homogeneous distribution of reinforcing particles into the molten matrix material and to increase wettability. As a result, choosing the right blade angle, diameter, and number of blade is critical for achieving a good amount of mixing.

Researchers employed a water model and a CFD model to evaluate the impact of impeller blade angle. The blade angles chosen were 15, 30, 45, 60 and 90 degrees. The influence of impeller blade angle on the distribution of solid particles in a liquid using a water model (Ravi et al., 2007). The results reveal that, a stirrer with a blade angle of 30 degrees operated well and produced uniform dispersion with no solid particles accumulation. Furthermore, stirring time has a significant impact on the dispersion of solid particles and the stirrer's power consumption. The usual image of a four-bladed stirrer is shown in the fig. 2.5.

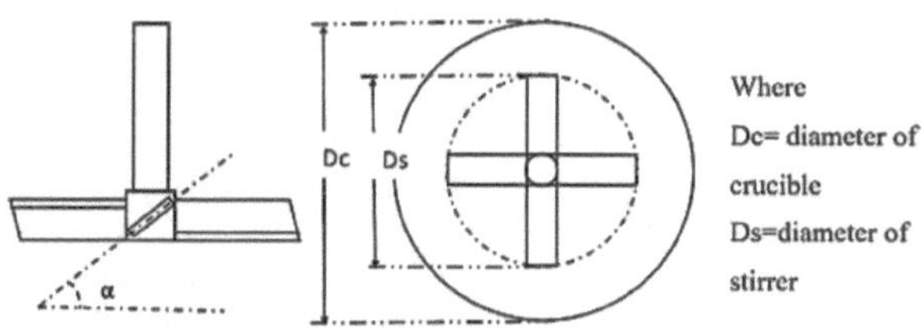

Fig. 2.5: Stirrer blade angle (Emiru, 2021).

For the optimization of stirring settings, researchers employed various models. The water model, the FEM model, and the experimental model. They tried to find the best settings for single-stage and multistage stirring type parameters impeller stirrer. To avoid clustering of reinforcement particles at the bottom, the blade should be less than or equal to 30% of the height of the fluid from the crucible's base. In addition, the blade's diameter should be half of the crucible's diameter. The position of the stirrer in the crucible during the activity is shown in fig. 2.6. The fishbone diagram in fig. 2.7 depicted the parameters that affect the quality of hybrid metal matrix composites produced by stir casting process.

h≤ Ho

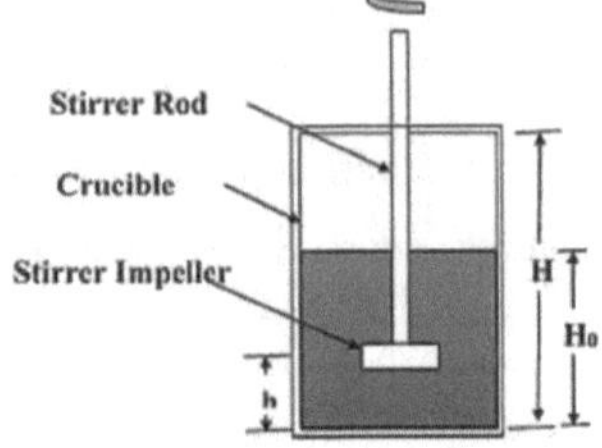

Fig. 2.6: Position of stirrer in the crucible (Ravi et al. 2007).

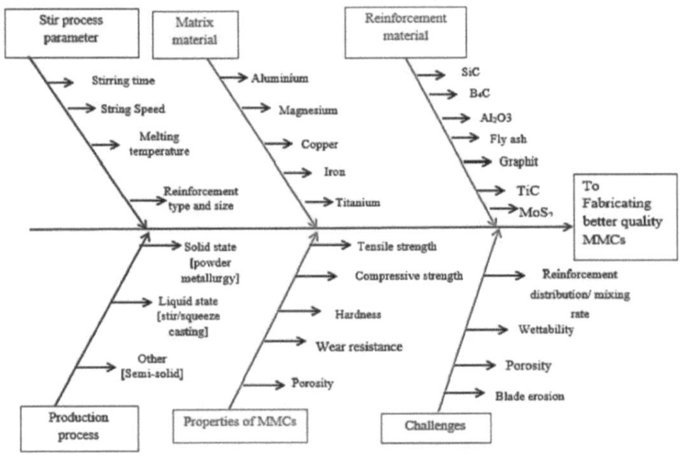

Fig. 2.7: Fish bone diagram for producing better quality MMCs by stir casting process (Emiru et al. 2021).

2.4.3 Modeling and Optimization of Process Parameters

The two most significant concerns in product manufacture are process modeling and optimization. A variety of dynamically interacting process factors characterize manufacturing operations. When it comes to machining, temperature is crucial. Thermal damage caused by high cutting temperatures causes geometrical inaccuracies in the produced product and shortens tool life. During the machining process, a significant portion of machine energy is converted to heat by plastic deformation of the work piece surface, chip friction on the tool face, and friction between the tool and the work piece. During the process of cutting metal with a machine tool, there are three main sources of heat generation (Nusrat Tarin Chowdhury, 2011):

- As the work piece undergoes massive irreversible plastic deformation, heat is generated in the primary shear zone.
- Heat generated on the tool rake face, or secondary shear zone, by friction and shear. The chip material is distorted further, and some of it sticks to the tool face.
- Heat is generated at the tool-work interface, when the tool flank rubs on the work piece surface and generates heat.

As the cutting action proceeds and the heat has been generated most of the heat is dissipated in the following manners (Nusrat Tarin Chowdhury, 2011):

- The heat is carried away by the thrown chip. Along the length of the chip, the temperature decreases.
- As a heat sink, the work piece is used.
- As a heat sink, the cutting tool is used.
- Coolant, if applied, will aid in the removal of heat from all locations.

Cutting force is an important machinability metric because it influences productivity, product quality, and overall machining cost. Cutting force often reduces as cutting speed increases. The surface roughness is known to be significantly affected by different cutting parameters like the depth of cut, spindle speed and feed rate. Surface roughness is also influenced by cutting parameters such as work material characteristics, work hardness, unstable built-up edge, cutting time, tool nose radius and tool cutting edge angles, machine tool and work piece setup stability, chatter, and the use of cutting fluids in variable degrees. As a result, if the proper cutting conditions are used, the surface roughness will be optimized.

Metal cutting optimization is determining the best set of operating parameters to achieve a cost-effective goal within the restrictions of the operation. An optimization problem entails optimizing one or more objective functions while adhering to a set of restrictions. These goals are frequently incompatible and incomparable. Increased rate of feeding leads to increased production rate, but it also raises the cost of the operation due to excessive tool wear and lowers the surface quality due to increased roughness.

With multiple well-designed experimental runs, Taguchi and Analysis of Variance (ANOVA) may easily optimize the cutting parameters. Taguchi parameter design can improve performance characteristics by adjusting design parameters and lowering sensitivity of system performance to source of variation (Nusrat Tarin Chowdhury, 2011).

The ant colony technique is used to solve multi-pass turning optimization problems. Roughing and finishing steps are included in the cutting process. Under various realistic machining limitations, the machining parameters are selected by reducing the unit production cost (Nusrat Tarin Chowdhury, 2011).

An optimal fuzzy logic controller design based on evolutionary algorithms, which are efficient and robust optimization techniques. It is demonstrated that the designed fuzzy logic controller can achieve automatic feed rate modification to optimize production rate in turning operations while maintaining a constant cutting force (Nusrat Tarin Chowdhury, 2011).

A number of writers employed geometric programming to calculate the best cutting speed and feed rate for single pass turning operations under a variety of limitations. Under diverse machining constraints, the created models and programs can be utilized to calculate the optimum cutting parameters that satisfy minimum production cost or maximum production rate in turning processes (Nusrat Tarin Chowdhury, 2011).

Simulated annealing (SA) and a genetic algorithm (GA) is used to find the best machining parameters for continuous profile machining at the lowest possible cost, while keeping a number of practical restrictions in mind. Cutting force, power constraint, and tool tip temperature are the restrictions in this problem (Nusrat Tarin Chowdhury, 2011).

Artificial neural networks are a new approach that provides efficient and quick selection of the best cutting conditions and processing of relevant technological data (ANN). A modified neural algorithm is used to optimize turning parameters, and their testing results demonstrate that the modified neural algorithm performs better in terms of maximizing production range (Nusrat Tarin Chowdhury, 2011).

A deterministic optimization approach involving mathematical analyses of constrained economic trends and graphical representation on the feed-speed domain results in a well-defined strategy that not only provides a unique global optimum solution, but also software suitable for on-line CAM applications. The established optimization algorithms and tools were validated by a numerical research, which also demonstrated the economic benefits of utilizing optimization (Nusrat Tarin Chowdhury, 2011).

RSM extracts the most information with the least amount of effort. It is a valuable tool for analyzing the impacts of several factors and their interactions on one or more response variables. Box and Wilson first proposed response surface methodology (RSM) in the early 1950s (Martland, 2013). RSM is a set of mathematical and statistical approaches that may be used to approximate and optimize stochastic models. The goal function associated with such models is referred to as noisy or stochastic objective function since it is prone to random disturbance.

2.5 Summary of the Literature

The literature review on metal matrix composite, its fabrication process and machining highlights the vast potential of the aluminum-based metal matrix composites. It is obvious that the reinforcements percentage on the matrix materials holds a vital influence on the characterization of the metal matrix composites. The effect of processing conditions and various fabrication methods (such as liquid state and solid state processing techniques) have been widely used.

The impact of weight composition reinforcement on the physical, mechanical, and tribological behavior of AMMCs has been investigated. In the manufacturing of aluminum matrix composites, ceramic particles were commonly used as reinforcements. According to the literature, adding various particle reinforcements to an aluminum alloy matrix improves the mechanical and tribological properties significantly. For tribological properties, there are many parameters that can be considered in optimizing machining operations. The few parameters reported to include depth of cut, feed rate, cutting speed, number of passes, tool diameter, tool length, and so on. The most three parameters used by the literature is the depth of cut, feed rate, and cutting speed.

Many works have so far been done to optimize the machining parameters by using different optimization techniques. But most of them face great difficulties when the number of variables increases because the problem becomes combinatorically explosive.

This research aims to develop hybrid particle reinforced aluminum-based metal matrix composites, observe the morphological, physical and mechanical properties of the developed composites and develop predictive models for drilling and optimize the machining parameters.

CHAPTER 3
MATERIAL DEVELOPMENT

3.1 Introduction

Major constituents in a metal matrix composite are reinforcements and the matrix. The matrix acts as a binder for the reinforcing agents. Similar to that, in a hybrid composite material also, reinforcements and the matrix are the main components. By impregnating two or more types of reinforcements in the same matrix, hybrid composites are formed. One reinforcements may be of low density, low hardness, low tensile strength and the other one may be of high density, high hardness, high tensile strength. Hybrid composites provide a large range of mechanical properties with a reasonably good price. However, to predict the mechanical properties of a hybrid composite the rate of mixture (ROM) may not work very well. They often show hybrid effect which is defined as the difference between the failure strain of high hardness reinforcement in a hybrid composite and that in a pure high hardness reinforced composites.

3.2 Selection of Materials

Choosing an eco-friendly lightweight, and high strength material begins with the composite of the material selecting the optimal reinforcements for use in composite materials is a difficult task. Due to the abundance of reinforcements available, each with a unique set of mechanical qualities, it becomes difficult to choose the best reinforcement in a clean manner. Thus, reinforcement selection for metal matrix composites involves comprehensive data analysis and statistical analysis to determine the optimal performance analysis.

3.2.1 Matrix Selection

Proposed by Kalpesh Maniya and M. G. Bhatt (2010) the PSI (preference selection index) method is one of the most popular statistical method for evaluating the order of preferences of alternative without deciding relative importance between attributes. This method is useful when there is conflict in deciding the relative importance between attributes. Using overall preference value, preference selection index (I_i) for each alternative is calculated and alterative with higher value of PSI is selected as best alternative.

The alternatives for this study were the twenty-two metals and alloy, namely iron, aluminum, nickel, beryllium, copper, magnesium, manganese, molybdenum, tin, titanium, tungsten, molybdenum alloy, cobalt alloy, aluminum alloy, beryllium alloy, magnesium alloy, nickel alloy, tin alloy, zinc alloy, tungsten alloy, alloy cast iron, titanium alloy.

Table 3.1 shows the data of alternatives.

Table 3.1: Decision matrix for alternative matrix materials

SN	Materials	Shear strength, MPA	Yield tensile strength, MPa	Hardness, Brinell	Young's modulus ,Gpa	Density, g/cc	Melting point, deg C	Thermal conductivity, W/m-K	Specific heat, J/g degc
1	Fe, iron	358	671	251	196	7.79	1440	16.7	0.478
2	Aluminum, AL	207	276	95	68.9	2.7	652	167	0.896
3	Nickel	723	590	282	207	8.4	1320	16.7	0.431
4	Beryllium	345	240	85	303	1.844	1283	216	1.925
5	Copper	210	33.3	50	110	7.764	1083	385	0.385
6	Mg, Magnesium	165	105	35	44	1.74	649.3	159	1.025
7	Manganese, Mn	496	241	460	159	7.44	1244	7.82	0.448
8	Molybdenum, Mo	500	324	225	330	10.22	2617	138	0.255
9	Tin, Sn	220	11	2.3	41.6	5.765	231.968	63.2	0.256
10	Titanium, Ti	220	140	70	116	4.5	1670	17	0.528
11	Tungsten, W	400	750	294	400	19.3	3370	163.3	0.134

Table 3.1: Continued

SN	Materials	Shear strength, MPA	Yield tensile strength, MPa	Hardness, Brinell	Young's modulus ,Gpa	Density, g/cc	Melting point, deg C	Thermal conductivity, W/m-K	Specific heat, J/g degc
12	Molybdenum alloy	500	567	181	303	10	2340	173	0.255
13	Cobalt alloy	415	760	323	191	8.4	1350	14.7	0.423
14	Aluminum alloy	165	180	84.5	72	2.73	579	135	0.962
15	Beryllium alloy	298	672	256	155	7.57	1010	150	1
16	Magnesium alloy	145	164	63.7	45.2	1.81	540	91.1	1.02
17	Nickel alloy	723	590	282	207	8.4	1320	16.7	0.431
18	Tin alloy	39.7	33	16.4	37.4	7.8	203	46.2	0.211
19	Zinc alloy	225	277	72.1	83.2	6.61	424	109	0.416
20	Tungsten alloy	394	678	925	340	15.4	2110	155	0.172
21	Alloy cast iron	712	520	262	156	7.17	1150	26.6	0.506
22	Titanium alloy, general	608	778	305	114	4.5	1650	7.64	0.519

After calculation of the preference selection index in appendix A, aluminum is selected as the matrix material. Fig. 3.1 shows the change in raking derived from PSI method.

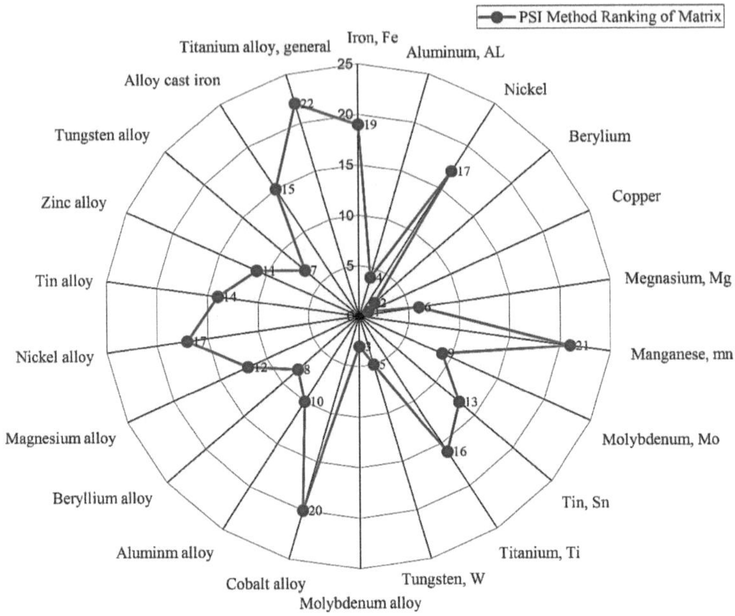

Fig. 3.1: Ranking of the material matrix derived from PSI analysis.

3.2.2 Reinforcement Selection

This study examines methods for assessing reinforcements performance. The performance scores of different reinforcements were evaluated using fuzzy TOPSIS (technique for order of preference by similarity to ideal solution) approach. Performance score is the reinforcements reaction variable to its attributes. Following that, fuzzy MOORA, and SAW analyses were carried out to determine the performance score and ranking of particulate reinforcements. The best alternative was assessed from these particulate reinforcements based on their performance score and ranking. Detailed steps demonstrated in the appendix B, appendix C, and appendix D. A sensitivity analysis was carried out with the new ranking of the particulate reinforcements derived by elimination the rank reversal phenomenon.

The Table 3.2 represents the decision matrix.

Table 3.2: Decision matrix for alternative particulate reinforcements

Serial No	Name	Density g/cc	Price dollar (2021)	Melting point deg C	Modulus of elasticity GPa
1	Aluminum oxide Al_2O_3	3.96	6	2054	370
2	Barium boride BaB_6	4.35	20	2270	385
3	Chromium carbide Cr_3C_2	6.67	0.66	1800	380
4	Hafnium Carbide HfC	12.2	105	3000	352
5	Niobium Carbide NbC	7.6	80	3500	450
6	Silicon Carbide SiC	3.1	4.5	2797	420
7	Tantalum Carbide TaC	14.3	228	3900	550
8	Titanium Carbide, TiC	4.94	20	3065	450
9	Vanadium Carbide VC	5.71	45	2730	350
10	Zirconium Carbide ZrC	6.65	89	3400	390
11	Chromium nitride CrN	5.9	15	1050	83
12	CNT	1.35	2	3550	800
13	GNP	2.26	50	3550	210
14	Boron carbide B4C	2.52	30	2350	289.65
15	Manganese dioxide MnO_2	5.026	50	535	276.34
16	Silicon nitride Si_3N_4	3.29	85	1000	310
17	Aluminum nitride AlN	3.2	9	1700	330

Table 3.2: Continued

Serial No	Name	Density g/cc	Price dollar (2021)	Melting point deg C	Modulus of elasticity GPa
18	Titanium Nitride TiN	5.22	15	2950	600
19	Zirconium Oxide ZrO_2	5.7	20	2750	205
20	Molybdenum Disulfide MoS_2	5.06	30	1185	330

With the criteria weights obtained from TOPSIS analysis were carried out for each reinforcements in appendix B. From table B.4, it is found carbon nanotube ranked first as the reinforcement, silicon carbide ranked second and aluminum oxide ranked third. These three particulate reinforcements were selected for developing hybrid particle reinforced metal matrix composites.

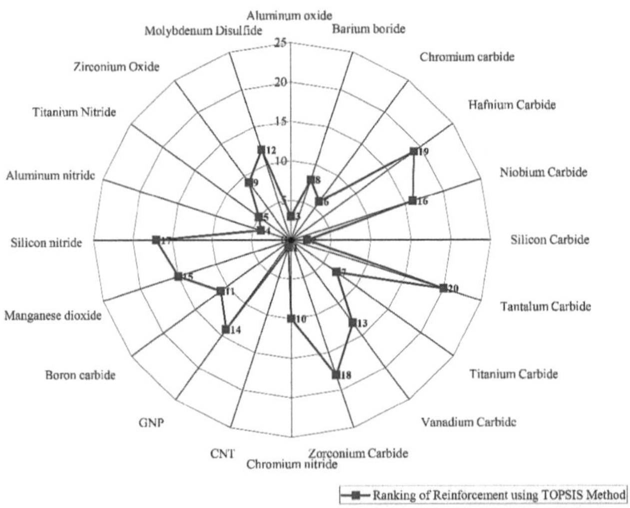

Fig. 3.2: Ranking of the reinforcement materials derived from fuzzy TOPSIS analysis.

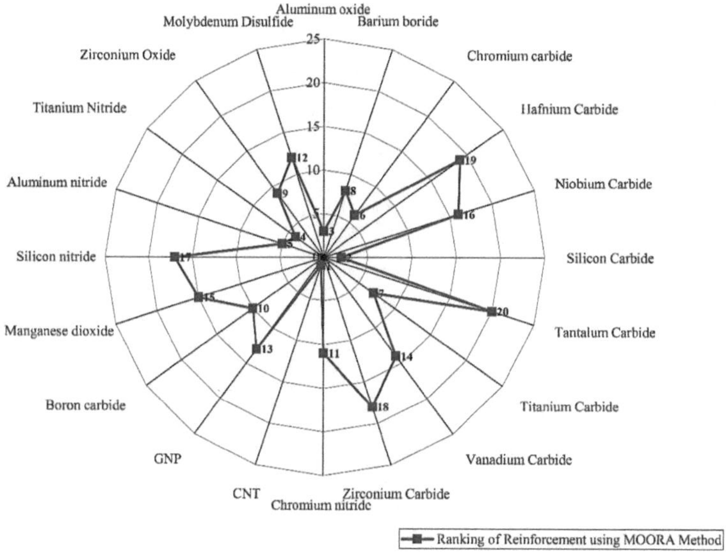

Fig. 3.3: Ranking of the reinforcement materials derived from fuzzy MOORA analysis.

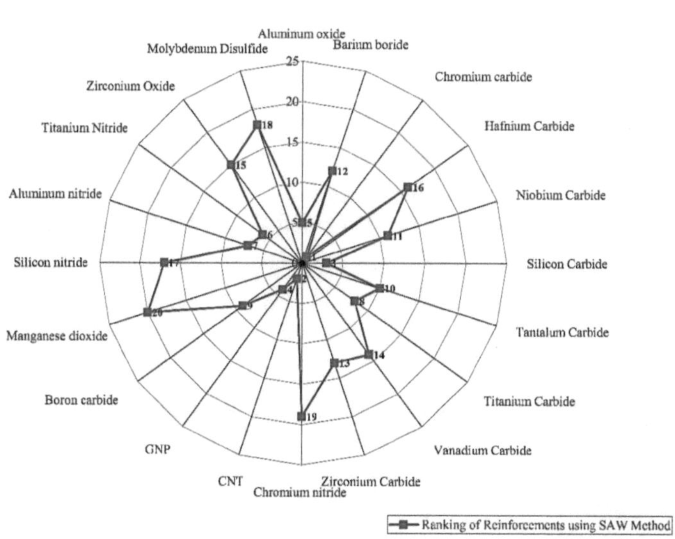

Fig. 3.4: Ranking of the reinforcement materials derived from fuzzy SAW analysis.

From Fig. 3.2 to Fig. 3.4 it found, the TOPSIS and MOORA offered almost identical order of preference, whereas SAW offered difference in ranking with the others. Carbon nanotube, alumina and silicon carbide were the close competitor for the top position in the order of preference of particulate reinforcements in the three methods. Fig 3.5 indicates the relationship among the ranking for the three methods.

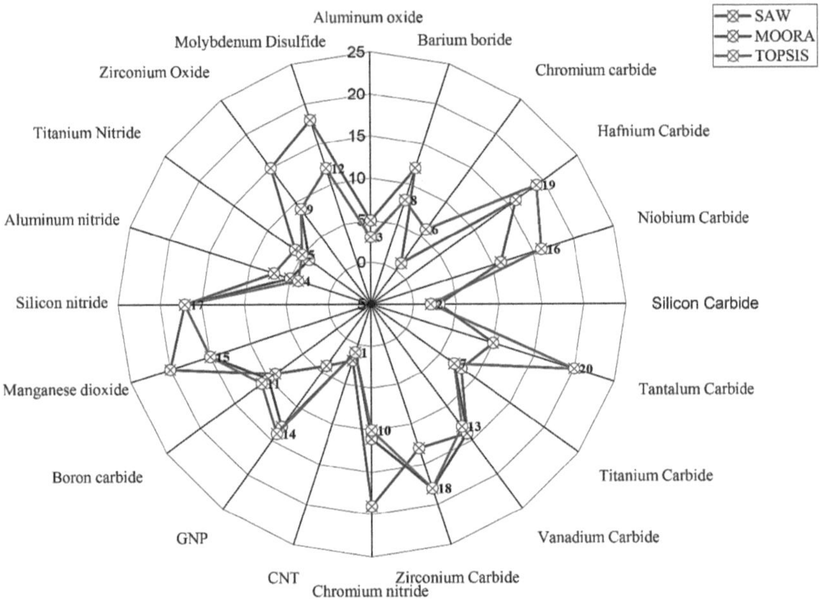

Fig. 3.5: Relationship among rankings obtained from different methods.

As expected, the top ranking was almost similar in the three methods, but no other specific correlation was observed between them. So, carbon nanotube, alumina and silicon carbide selected to fabricate the hybrid particulate reinforced metal matrix composites.

Table 3.3: Selection of alternative from fuzzy TOPSIS, MOORA and SAW methods

	SAW-MOORA	SAW-TOPSIS	MOORA-TOPSIS
Spearman correlation coefficient	0.656	0.624	0.995

Table 3.3 pinpoints the correlation between the rankings from three methods by evaluating the Spearman correlation coefficient. Positive correlation found for all the cases SAW-MOORA, SAW-TOPSIS and MOORA-TOPSIS analysis, where the last couple has the most correlation between each other. It indicates the suitability for MOORA-TOPSIS method, whereas, for SAW-MOORA and SAW-TOPSIS methods null hypothesis is highly applicable.

3.3 Metal Matrix Composite Fabrication Process

In metal matrix composite (MMC) fabrication process stir casting process was used. A mechanical stirrer is required to form a vortex for the necessity of the uniform distribution of reinforcing particles into the molten matrix material. Therefore, proper selection of a suitable blade angle, diameter, and number is crucial to obtain a good level of mixing. In this work, for the fabrication of a simple stirrer setup, 3mm thick sheet metal, 13mm diameter mild steel circular rod, 4mm thick mild steel, and hand drill machine with approximate 300RPM is used. Before the fabrication of the stirrer setup, the dimension of the furnace and the dimension of the crucible were measured properly, and the stirrer setup fabricated accordingly.

Sand casting process is the most widely process in the manufacturing sector due to its various advantages over other manufacturing processes. Sand casting is resistant to elevated temperature, almost all metal can be cast using the sand casting process.

The size of the mold required to fabricate the composite depends on the size of specimens used for various tests. All specimens for various tests were fabricated at once using the molds.

To make superior-quality single or hybrid particle reinforced metal matrix composite materials, the stir or squeezed-stir casting technique is extensively used. The melting temperature of the matrix, die temperature, stirring speed, stirring time, reinforcement weight percent, reinforcement preheating temperature, and other process variables all play a role in the manufacture of high-quality metal matrix composites in the stirring casting process.

The fish bone diagram in Fig. 3.6 shows the fabrication steps for stir casting procedure.

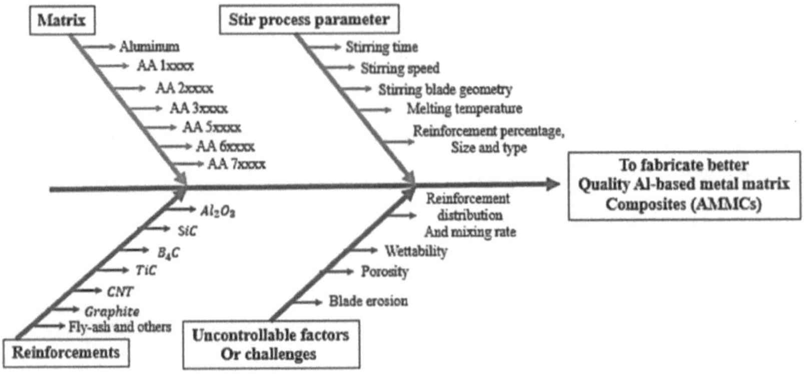

Fig. 3.6: Fish bone diagram for producing better quality AMMCs by stir casting process.

3.4 MMC Material Development

Matrix is a component of the composite materials that is used to bind reinforcements together using its cohesive and adhesive characteristics, transfer load to and between reinforcements, and protect reinforcements from environmental factors such as humidity, and chemical attack. MMCs are the most widely used composites in the industrial sector. The aluminum alloy is used as the matrix in this work.

Aluminum alloy is employed as the working matrix material for the manufacturing of hybrid metal matrix composites because of its properties and availability. Fig 3.7 depicts aluminum alloy scraps used as the matrix material in the fabrication of hybrid particle reinforced metal matrix composites. We cannot determine the quality of the material since it was sourced from different suppliers.

Fig. 3.7: Aluminum alloy scrap for composite fabrication.

The function of the reinforcements is extremely important since they are responsible for carrying the significant percentage of the applied load. Keeping this in thought, in this research, three different reinforcements namely carbon nanotube, silicon carbide and aluminum oxide have been used as the reinforcing material for the developed hybrid particle reinforced composite.

Table 3.4 lists the materials used in this study to develop a hybrid particle reinforced aluminum based metal matrix composite, including aluminum alloy as the base material and silicon carbide, aluminum oxide and carbon nanotube as reinforcement materials, as well as the amount of each material required in kg.

Table 3.4: Materials required in Kg to fabricate hybrid particle reinforced aluminium metal matrix composites

Serial No.	Items	Quantity	Wt. Percentage
1	Aluminium alloy	25kg	94 %
2	Silicon Carbide (SiC)	500gm	2.5%
3	Aluminum oxide (Al_2O_3)	500gm	2.5%
4	Carbon Nanotube (CNT)	250gm	1%

Composite properties are influenced by the weight percentage (wt.%) of reinforcing materials in the matrix material. Aside from the primary material qualities, the geometry of the reinforcement (form, size, and distribution) has a significant impact on the composites qualities. The orientation of the reinforcement has an impact on the composite characteristic.

Table 3.5 to 3.7 illustrates the properties of the carbon nanotube, silicon carbide and aluminum oxide particulate reinforcements.

Carbon nanotubes are one of the most fascinating nanomaterial discoveries of the twentieth century. Since Sumio Jijima developed CNT in 1991, it has been regarded as one of the most ideal reinforcing materials for metal matrix composites. Shi-ying et al. 2010, looked into the microstructural and mechanical features of the materials. AZ91D magnesium metal matrix composite enhanced with carbon nanotubes composites were created using mechanical stirring and a high-intensity ultrasonic dispersion method. The results reveal that CNT is spread evenly throughout the metal matrix. Tensile strength, yield strength, and elongation of composites reinforced with 1.5% CNT are 22%, 21% and 42% higher than AZ91D matrix alloy. In metal matrix composites, CNT can also help to prevent local crack propagation to some extent. Zhou et al. 2018, used powder metallurgy technique to fabricate CNT reinforced AZ31 metal matrix composites and investigated the effect of CNT on the mechanical and wear properties of the alloy. In comparison to AZ31 alloys, the yield strength, tensile strength, and micro hardness of the composite increased. The addition of CNT may weaken the basal plane, reducing the strengthening effect of the material. Due to the self-lubricating effect of CNTs and the creation of carbon film covering the wear surface, adding CNTs to the metal matrix can greatly lower the friction coefficient and weight loss of the matrix.

Table 3.5: Properties of carbon nanotube

Density (g/cc)	Average diameter, mm	Average length, mm	Purity	Surface area, m^2/g	Poisson's ratio
1.4	30-80	<20	>99.0%	80-120	0.27

Silicon carbide particles have grown in popularity as a popular reinforcement material many aluminium and magnesium alloy-based material metal matrix composites. They are hard and brittle with high strength, high modulus of elasticity and high thermal resistance. The size of the SiC particles depends on the manufacturer and the type of alloy. However, the mean particle size lies in the range of 2-20μm. Deng et al. 2010, investigated the microstructural and mechanical properties of Mg MMCs reinforced with SiC at a higher temperature with performing forging operation to reduce the height of the sample. The MMCs prepared by stir casting process and adding reinforcement particulates in Mg ally AZ91. The texture analyses of forged composites carried out by neutron diffraction. Silicon carbide is chosen as a reinforcement material because it has exclusive properties such as high hardness and high toughness, thermal and chemical resistance etc.

Table 3.6: Properties of silicon carbide

Density (g/cc)	Melting point deg C	Form	Purity	Particle size, mesh	Poisson's ratio	Hardness, Kg/mm^2	Young's modulus, GPa
3.21	2730	Powder	97.0%	50-220	0.27	2800	410

Alumina particles are being used more and more often for research and industrial application, due to their improved properties compared to bulk materials. The benefits of nano particles can enhanced in toughness and ductility, also increased hardness and strength of metals and its alloy. Park et al. 2008, studied the effect of alumina in aluminium metal matrix composite by varying their volume fraction of alumina from 5 to 30 volume percent. The results showed that by increasing the volume fraction of alumina, fracture toughness of Al MMCs decreased. This is due to inter particles spacing between nucleated micro voids.

Table 3.7: Properties of alumina

Density (g/cc)	Melting point deg C	Form	Purity	Particle size, mesh	Poisson's ratio	Hardness, Kg/mm^2	Young's modulus, GPa
3.96	2054	Powder	98.5%	63-220	0.22	5500	370

To make superior-quality single or hybrid particle reinforced metal matrix composite materials, the stir or squeezed-stir casting technique is extensively used. The melting temperature of the matrix, die temperature, stirring speed, stirring time, reinforcement weight percent, reinforcement preheating temperature, and other process variables all play a role in the manufacture of high-quality metal matrix composites in the stirring casting process.

Sand preparation is one of the sophisticated process in the sand casting procedure, shown in Fig. 3.8. The sand gives the mold shape for the geometry. To prepare the sand, 7% sodium silicate and wood coir was mixed with the sand and mixed in the sand mixture machine. After the sand mixed, the sand was placed in the prepared mold box and carbon dioxide gas blow through the box. The gas reacts with the sodium silicate mixed with the sand earlier and create a compact bonding and smooth surface for the pattern (Kumar et al. 2021). Then by using the shape of the wooden pattern create cavity created and mold coat eleven was used to coat the cavity. It coating helps to avoid moisture creation and delamination of sands.

Fig. 3.8: Sand preparation using pattern.

According to the furnace's dimensional specifications, a graphite crucible with a capacity of 40kg was chosen. The crucible was warmed for two hours shown in Fig. 3.9 and Fig. 3.10, to eliminate moisture from the surface and to assist the crucible resist the initial shock when the matrix aluminum alloy metal was put on top of it. The base material aluminum was heated to 750 degrees Celsius within the graphite crucible to achieve the molten state, which is higher than the melting point temperature of aluminum alloy (Sharma et al. 2019).

Fig. 3.9: Furnace and crucible preparation.

Fig. 3.10: Material preheating.

When the crucible is filled with melted aluminum, carbon nanotube, silicon carbide and aluminum oxide reinforcements particles are placed in separate vessels and preheated beside the furnace. After obtain a desired preheating temperature, the reinforcements were added in the melted aluminum alloy maintaining the desired ratio. After that, the composition was stirred using a custom made stirring system. In the stir casting configuration, a mechanical stirrer is required to produce a vortex for the uniform dispersion of reinforcing particles into the molten matrix material and to improve wettability. As a result, selecting the appropriate blade angle, diameter, and number of blades is crucial for obtaining adequate mixing. Researchers used a variety of models to optimize the stirring settings. The water model, the FEM model, and the experimental model are all examples of models. They attempted to identify the ideal settings for single stage and multistage impeller stirrer stirring type characteristics. The blade should be less than or equal to 30% of the fluid's height from the crucible's base to avoid gathering of reinforcing particles at the bottom. Furthermore, the blade's diameter should be half that of the crucible (Premnath et al. 2014).

Fig. 3.11: Stirring blade setup (before Boron nitride) coating.

$h \leq H_O$

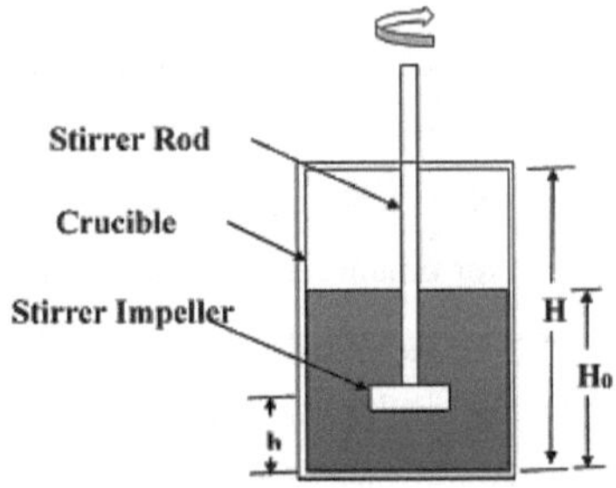

Fig. 3.12: Position of stirrer in the crucible (Premnath et al. 2014).

The stirring action is done for 15 minutes at approximate 250-300 rpm variation. Before starting the stirring, the impeller blades were coated with boron nitride create an insulation system and avoiding the blade from preheating. Fig. 3.11 to 3.13 shows the stirring setup for mixing the aluminum based metal matrix composite.

Fig. 3.13: Stirring of the molten composite.

The development of dissolved gases from melting, as well as slag containing gas porosity, are the main causes of gas porosity in castings. Although hydrogen may dissipate in liquid aluminum, its fluidity in solid aluminum is substantially lower, resulting in porosity. To avoid porosity development, a degasser was utilized in the melted material once the stirring operation was done. The molten mixture is then poured into sand molds and placed in a secure location to cool for 24 hours shown in Fig. 3.14.

Fig. 3.14: Materials pouring and specimen preparation.

Following the successful fabrication of the composite material using the stir casting process, the composite was prepared for physical, mechanical testing and microstructural characterization as a specimen. The specimen is produced to the desired form and size according to ASTM standard by machining on a milling machine for physical mechanical, and tribological tests.

Fig. 3.15: Milling operation of prepared composite.

Figure 3.15 shows the machining of a composite specimen on a milling machine. The main object of Grinding and Polishing of specimens is to generate a flat surface; removes cut marks; levels and free from excessive deformation. Some of the steps are; rough grinding, fine grinding, and polishing.

CHAPTER 4
EXPERIMENTAL PROCEDURE

4.1 Introduction

After successfully fabricate the composite materials using stir casting method, the composite was then prepared for a specimen of a physical test, morphological test, mechanical test and tribological test. For physical, mechanical and tribological tests, the specimens are prepared to the required shape and size according to ASTM standard by machining on milling. Chemical enchant was also used to prepare the specimens for microstructural tests.

4.2 Morphological Testing

To observe morphological characteristics optical microscopy and scanning electron microscope study was performed. A 50x50x10 mm^3 specimen polished according to ASTM standards was used for microstructural analysis.

In the first step, cut specimens are ground with sandpaper of grit sizes of 120, 400, 800, 1200, and 2000 grits. After grinding, to removes the artifacts of little stock and make the surface mirror, polishing was performed by adding diamond suspension.

For observing the optical microstructure, to etch the surface of the sample, a solution is used shown in Table 4.1. Fig. 4.1 shows the polishing machine to prepare the specimens for the morphological test and Fig. 4.2 shows the optical microscopic setup to observed microstructure. Fig. 4.3 represent the scanning electron microscope (SEM) setup to capture the SEM microstructure and X-ray diffraction peaks.

Table 4.1: Keller's etchant for aluminum alloy (Senthil et al., 2019)

Keller's etchant	Concentration (mL)	Conditions
Distilled water	92	10-30 second immersion (use fresh)
Nitric acid	6	
Hydrochloric acid	2	

Fig. 4.1: Polishing machine to prepare the composite for etching (NANO 2000T Grinder-Polisher).

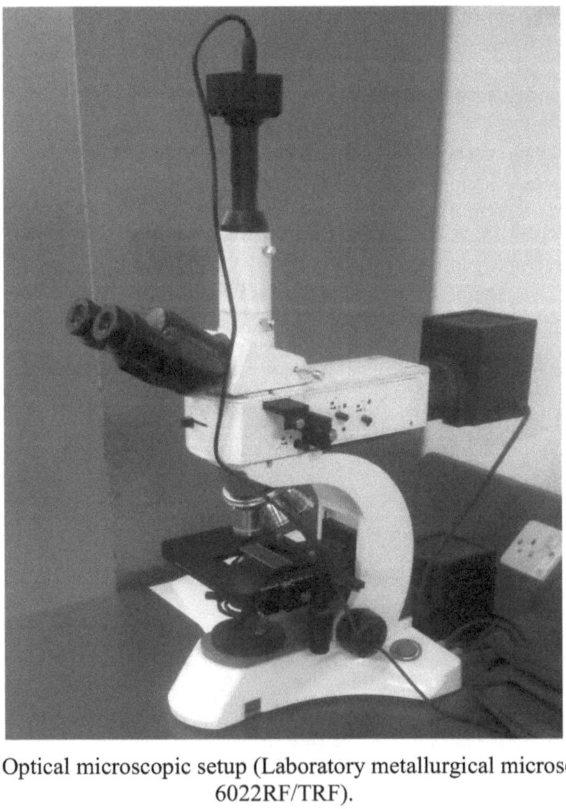
Fig. 4.2: Optical microscopic setup (Laboratory metallurgical microscope BS-6022RF/TRF).

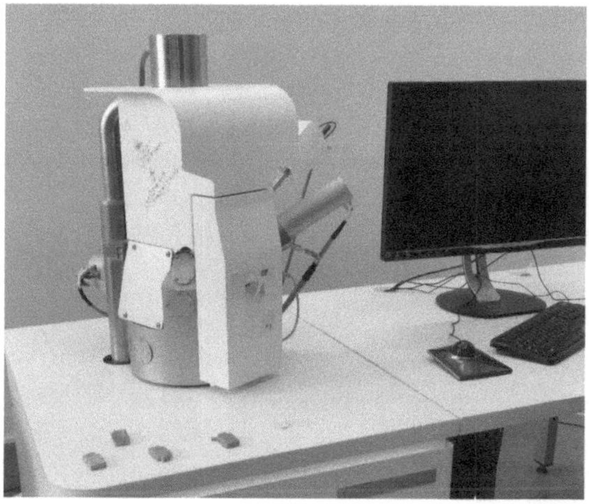

Fig. 4.3: Scanning electron microscope setup (Tescan Vega).

4.3 Density and Porosity Measurement

Density and porosity are the physical property of the material. Theoretically, the density of the material generally can be defined as mass over unit volume. A 50x50x10 mm^3 specimen is prepared to calculate the experimental density of the fabricated composites.

The theoretical density ρ_{th} of hybrid particle reinforced aluminum matrix composite was calculated using the rule of mixtures as shown in Eq 4.1 (Rajesh et al., 2018). Consider, the produced composites are the mixture of aluminum alloy, alumina, carbon nanotube, silicon carbide.

$$\frac{1}{\rho_{thx}} = \frac{wt\%_{Al}}{\rho_{Al}} + \frac{wt\%_{Al_2O_3}}{\rho_{Al_2O_3}} + \frac{wt\%_{SiC}}{\rho_{SiC}} + \frac{wt\%_{CNT}}{\rho_{CNT}} \tag{4.1}$$

Where, ρ_{thx} is the theoretical density of composite specimen x= 1,2,...n; $wt\%_{Al}$ is the mass percent of aluminum alloy in the composite specimen; $wt\%_{Al_2O_3}$ is the mass percent of alumina in the composite specimen; $wt\%_{SiC}$ is the mass percent of silicon carbide in the composite specimen; $wt\%_{CNT}$ is the mass percent of the carbon nano tube in the specimen; ρ_{Al} is the density of the aluminum; $\rho_{Al_2O_3}$ is the density of alumina; ρ_{SiC} is the density of silicon carbide; ρ_{CNT} is the density of carbon nanotube.

Experimentally, the density of hybrid particle reinforced aluminum matrix composite can be formulated as follows:

The actual density of composite can be calculated using volume, mass and density equation. The porosity of each composite material can be determined based on Eq. 4.2. (Aatthisugan, 2017).

$$P = 1 - (\frac{\rho_a}{\rho_{th}}) \quad (4.2)$$

Where, P is the percent porosity; ρ_a is the actual density of the composite material and ρ_{th} is the theoretical density of the composite material.

4.4 Mechanical Properties Measurement

Tensile strength of a material is the ability of a material to resist loads which applied in the longitudinal direction of specimen in out ward direction. Tensile specimens are prepared according to ASTM standards shown in Fig. 4.4 to Fig 4.6.

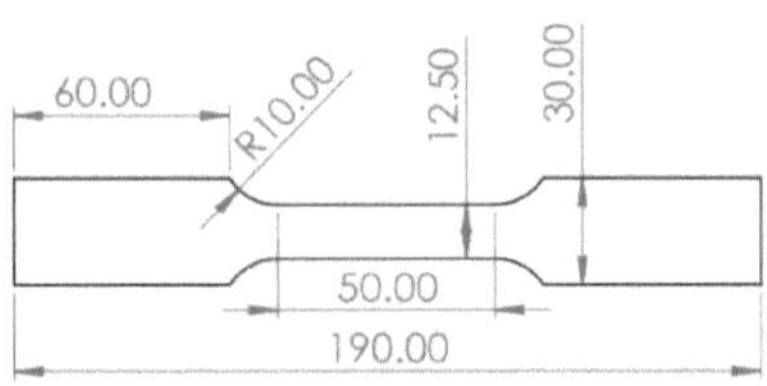

Fig. 4.4: ASTM E-8 standard tensile specimen.

Fig. 4.5: Tensile test specimens.

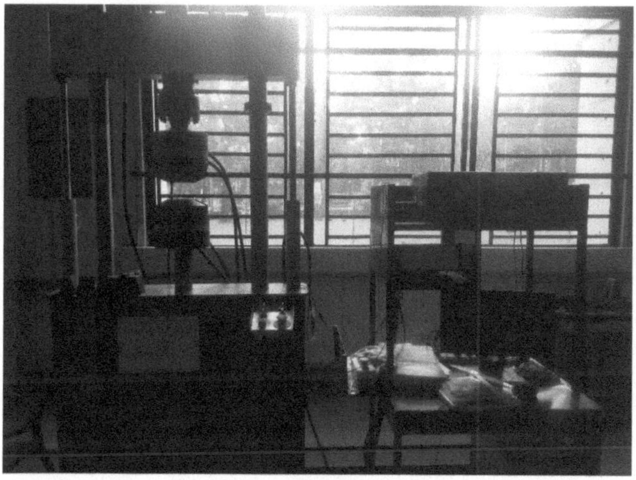

Fig. 4.6: Setup for tensile testing and flexural rigidity testing (Universal tensile tester PLS100).

Flexural strength is defined as the force couple required to bend a fixed non-rigid structure by one unit of curvature, or as the resistance offered by a structure while undergoing bending. Flexural strength measurement specimens are prepared according to ASTM standards shown in Fig. 4.7.

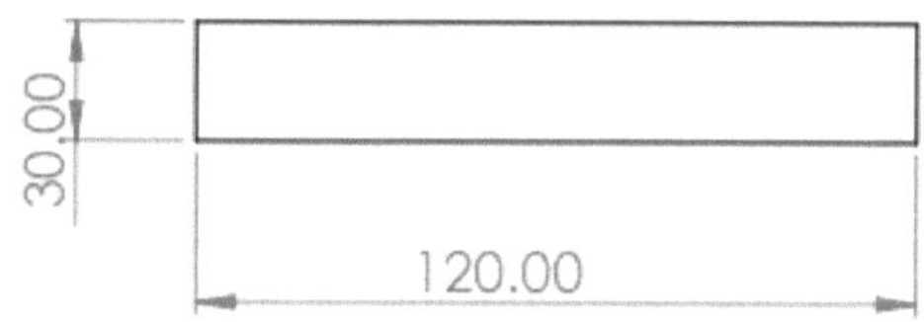

Fig. 4.7: Standard flexural rigidity testing specimen dimension.

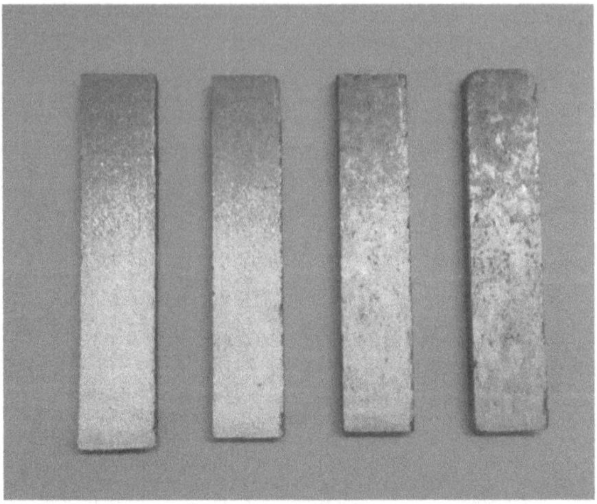

Fig. 4.8: Specimens for flexural rigidity testing.

In present work, Brinell hardness test machine was used to measure the hardness value of the fabricated hybrid particle reinforced aluminum metal matrix composites. Before hardness test, the surface of the material is grind and polished. Fig. 4.10 shows the specimens used for hardness test.

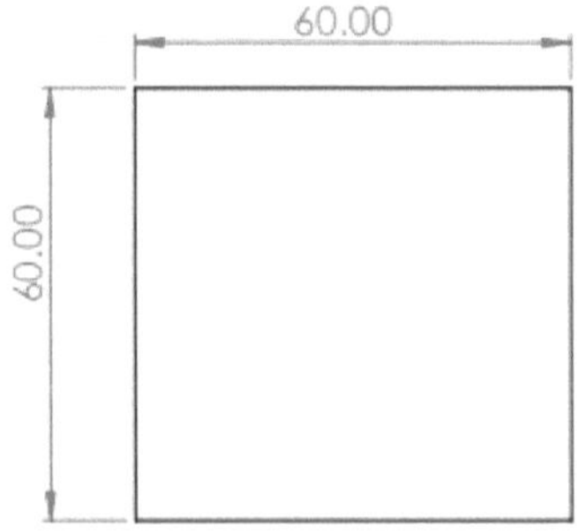

Fig. 4.9: Standard hardness test specimens dimension.

Fig. 4.10: Hardness test specimens.

Fig. 4.11: Setup for hardness testing (Brooks hardness tester, RAS).

The impact test is a standardized high strain rate test which determines the amount of energy absorbed by a material during fracture. Impact specimens are prepared according to ASTM standards shown in fig. 4.13 and the test setup shown in fig. 4.14.

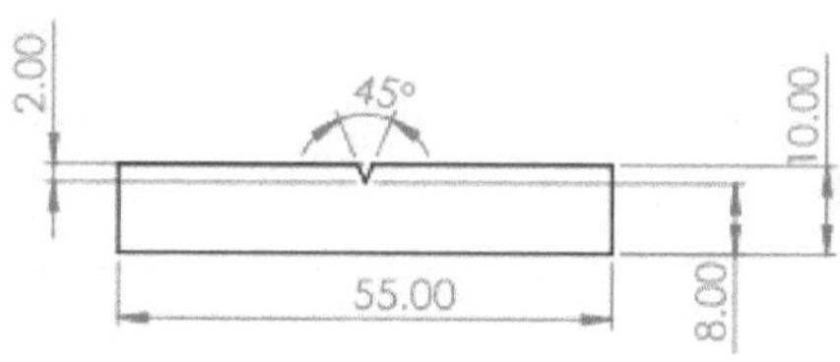

Fig. 4.12: Standard impact test specimens dimension.

Fig. 4.13: Impact testing specimens.

Fig. 4.14: Setup for impact testing (Universal impact tester, AIT300kN).

4.5 Machinability Test

To perform the machinability test, universal drilling machine is used. Fig. 4.15(a) illustrates the drilling experiment setup. Rotational speed (400, 1140 and 2260) rpm, and feed rate (0.1, 0.15, 0.2) mm/rev adopted as standard parameters in Sunlike Machinery (Model No: SN-Z3050x16) to keep less drilling-induced damages. 9 holes were drilled in dry environment in each specimen as shown in Fig. 4.15(c). During the drilling process chips were formed and for continuous engagement between work and tool, maximum heat was carried away through the chips and the rest stayed with the tool and work. Drilling condition of the experiment is shown in Table 4.2.

(a) (b) (c)

Fig. 4.15: Machining process, a) HMMC specimen work-piece, b) Experimental setup, c) Drilled HMMC specimen.

Table 4.2: Drilling condition of the experiment

Machine type	Machine model	Work material	Cutting tool	Rotation speed (rpm)	Feed rate (mm/rev)	Environment
Radial drilling machine	Sunlike Machinery (Model no: SN-23050x16)	1 Al-alloy and 3 AMMCs	Twist HSS drill (Ø=8mm)	400 ,1140 and 2260	0.1, 0.15 and 0.2	Dry

CHAPTER 5
RESULTS AND DISCUSSIONS

5.1 Introduction

Experimental investigation of aluminum-based metal matrix composites (MMCs) involves studying the properties and behavior of these materials under different conditions. The experimental investigation involves conducting various morphological, mechanical, and tribological tests to determine their properties, including microscopic tests, scanning electron microscopy (SEM), tensile strength, flexural strength, impact strength, hardness, and machining properties of the developed material. The results of these tests provide valuable insights into the performance of the developed aluminum-based MMCs and can be used to optimize their composition and processing for specific applications.

Contrary to pure materials and alloys, composite materials are inhomogeneous and anisotropic in nature. Their behaviors depend upon factors such as the reinforcement and matrix properties, the orientation of the reinforcements, the bond strength between the reinforcements and the matrix, and the type of fabrication process that follows to manufacture the composite. Taking these factors into consideration, in order to characterize the morphological property microscopic and SEM tests were done; to characterize the mechanical properties of tensile strength, flexural strength, impact strength, and hardness tests were done; to characterize the machinability properties of the developed work material, drilling and turning operation was done on the workpiece using different combination cutting speed, feed, depth of cut at dry machining environment.

The differences in physical, chemical, and mechanical properties between aluminum alloy and composite materials have led us to perform morphological, mechanical, and machining investigations on the developed composite materials.

5.2 Morphological Characterization

From the morphological investigation, it can be observing that the dispersion of particles in the matrix properly. Fig. 5.1 to Fig. 5.4 shows the microstructure of the composites at un-etched conditions.

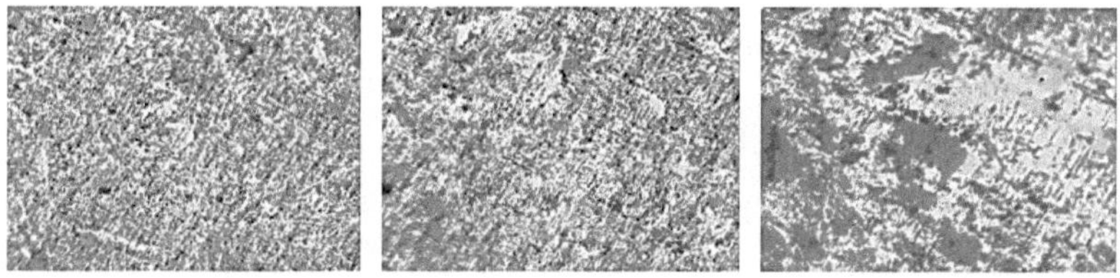

(a) (b) (c)

Fig. 5.1: Microstructure of Al-alloy (un-etched), a) 160x magnification factor, b) 320x magnification factor, c) 1600x magnification factor.

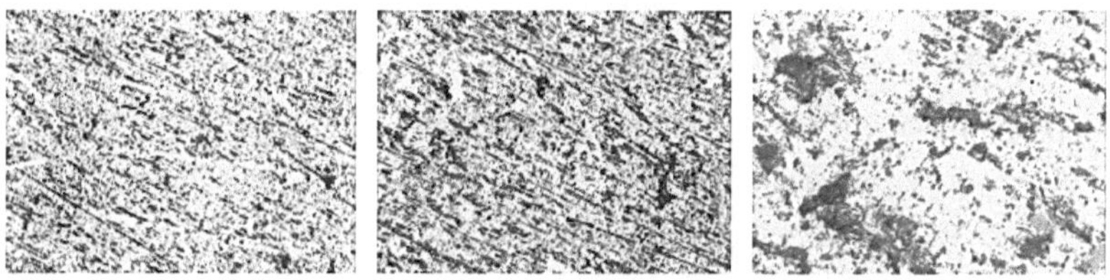

(a) (b) (c)

Fig. 5.2: Microstructure of composite of Al-alloy and CNT (un-etched), a) 160x magnification factor, b) 320x magnification factor, c) 1600x magnification factor.

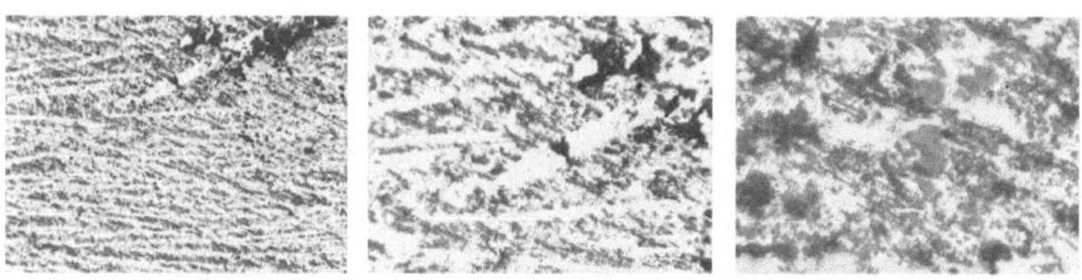

(a) (b) (c)

Fig. 5.3: Microstructure of composite of Al-alloy, CNT, and silicon carbide (un-etched), a) 160x magnification factor, b) 320x magnification factor, c) 1600x magnification factor.

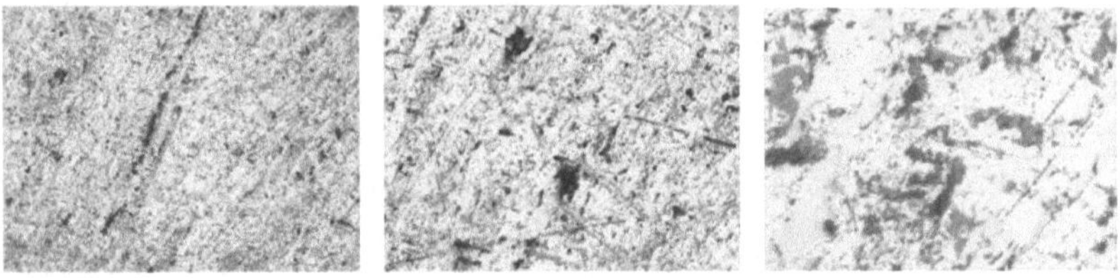

(a) (b) (c)

Fig. 5.4: Microstructure of composite of Al-alloy, CNT, silicon carbide, and alumina (un-etched), a) 160x magnification factor, b) 320x magnification factor, c) 1600x magnification factor.

From the figures, we observe the scars of machining and scattered particles reinforcements on the surface of the composites. To get a better view of particle dispersion in the composite, an optical micrograph at 160x magnification, 320x magnification, and 1600x magnification was obtained. The observation from the micrographs gives an understanding of phase distribution, and reinforcement interfacial bonding features with the aluminum alloy, as the properties of matrix composites depend on these attributes. A micrograph of fabricated metal matrix composites is presented in Fig. 5.5 to Fig. 5.8. Dispersion of carbon nanotubes, silicon carbide, and alumina particles observed from the figures. The optical micrograph at 1600x magnification factor is not very clear but a better understanding of particle size can be obtained.

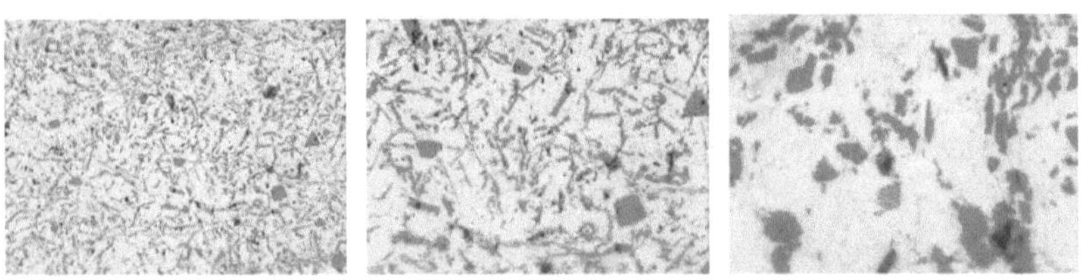

(a) (b) (c)

Fig. 5.5: Microstructure of composite of Al-alloy (etched), a) 160x magnification factor, b) 320x magnification factor, c) 1600x magnification factor.

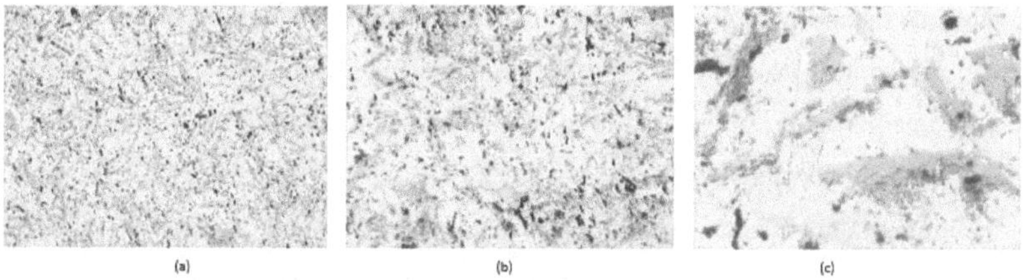

Fig. 5.6: Microstructure of composite of Al-alloy and CNT (etched), a) 160x magnification factor, b) 320x magnification factor, c) 1600x magnification factor.

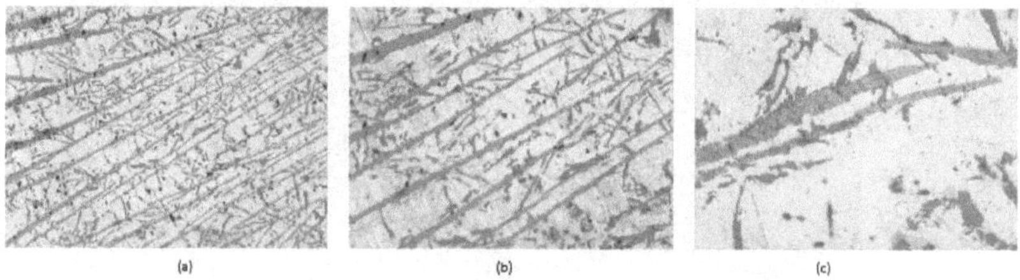

Fig. 5.7: Microstructure of composite of Al-alloy, CNT, and silicon carbide (etched), a) 160x magnification factor, b) 320x magnification factor, c) 1600x magnification factor.

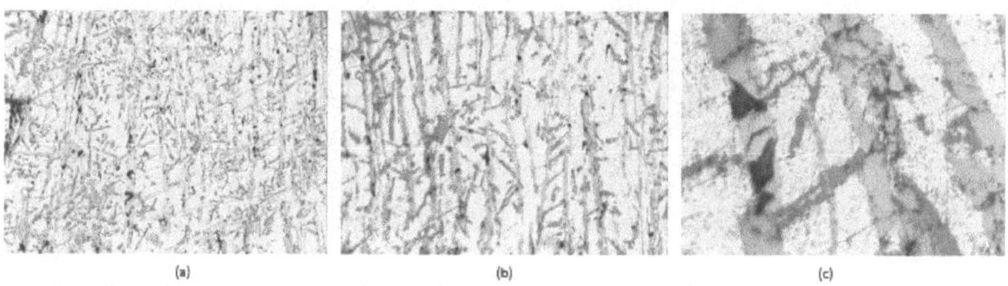

Fig. 5.8: Microstructure of composite of Al-alloy, CNT, silicon carbide, and alumina (etched), a) 160x magnification factor, b) 320x magnification factor, c) 1600x magnification factor.

From the optical micrographs of the composites, large interconnected coarse grains made up of intermetallic array were observed for the unreinforced aluminum metal matrix. The precipitated coarse grains affect the enhancement of properties; as minimum mechanical properties performance was obtained by the unreinforced aluminum alloy. It could generally be observed from the composites with silicon carbide, alumina and carbon nanotube reinforcements on aluminum alloy visibly changed the morphological structure. The influence of the reinforcement in improving the mechanical characteristics differed according to the percentage ration of the reinforcements. The microstructural results revealed a homogenous distribution of the reinforcement particles in the matrix phase. The surface morphology of the aluminum metal matrix composites was further investigated by obtaining SEM micrographs. The carbon nanotubes show an obviously bonded formation, with varied sizes and shapes. This formation is also observed by other researchers (Bharat et al. 2020).

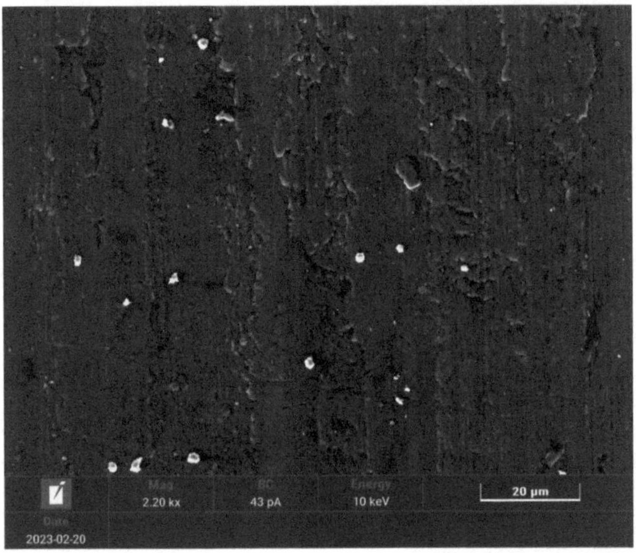

Fig. 5.9: SEM Microstructure of the composite of Al-alloy at 20µm magnification.

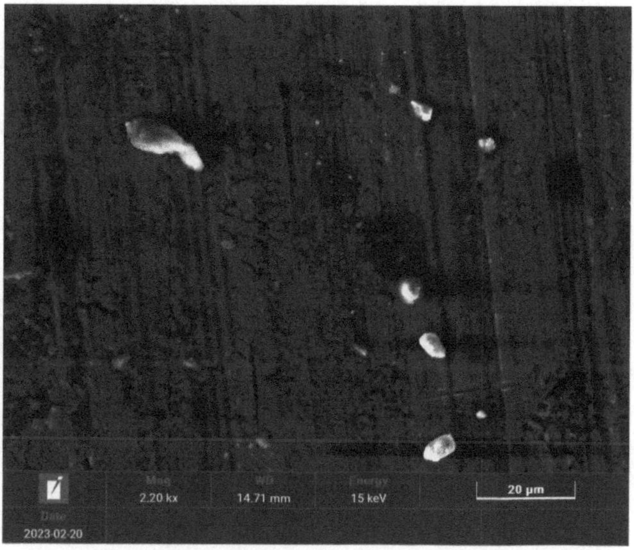

Fig. 5.10: SEM Microstructure of the composite of Al-alloy and CNT at 20μm magnification.

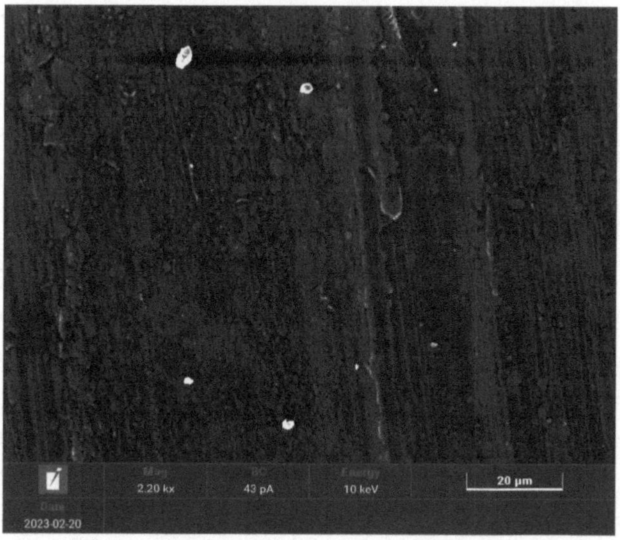

Fig. 5.11: SEM Microstructure of the composite of Al-alloy, CNT, and Silicon-carbide at 20μm magnification.

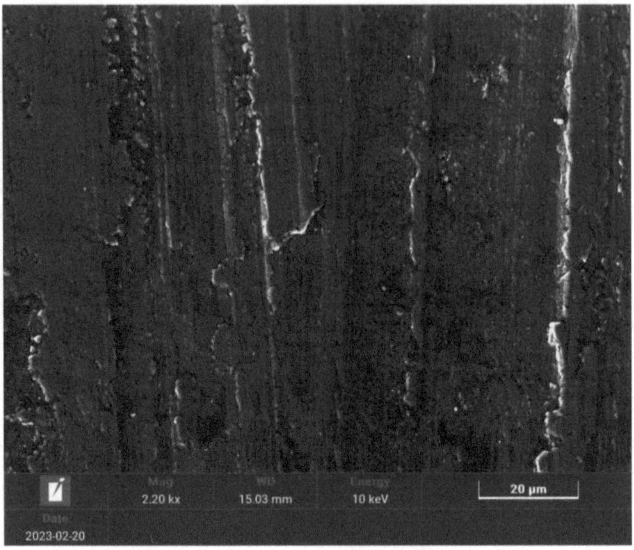

Figure 5.12 SEM Microstructure of the composite of Al-alloy, CNT, Silicon-carbide, and Alumina at 20µm magnification.

Through SEM images in Fig 5.9 to Fig. 5.12, it can be seen that distribution of alumina, silicon carbide and carbon nanotube reinforcements in aluminum matrix is very much fair. For the alloy, we also observed calcium and sodium presence in the metal. Also, it can be seen that there is little cluster of carbon nanotube in the hybrid particle reinforced composites and porosity in fabricated composites. Hence confirming the success of stir casting method to fabricate metal matrix composite. Uniformly spread particles with good wettability increases the strength of the composites produced.

Energy dispersion spectroscopic analysis is a non-destructive testing method of discloses the chemical constituents present in the cast composite by measuring the variation of the energy levels within the sample and is obtained as a result of focusing the electron beam. Figure 5.13 to 5.16 shows the spectroscopic images of aluminum alloy and the composites. It is evident from the spectral images that the peaks corresponding to aluminum is of maximum heights, as well as the peaks of carbon, oxide, silicon and carbide confirms the composition of hybrid particle in the base metal matrix.

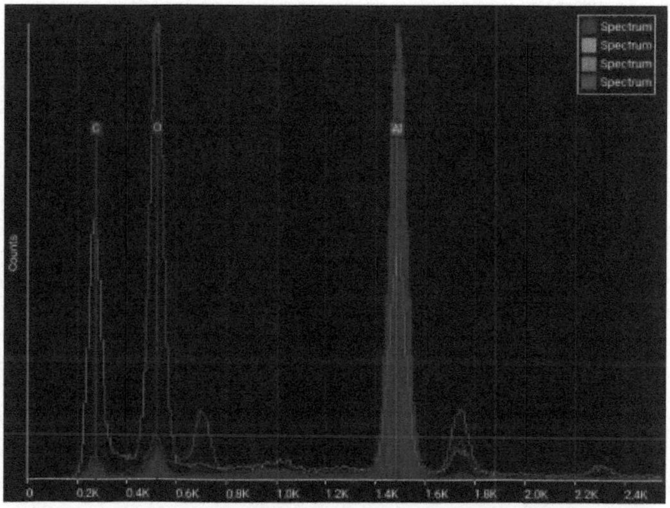

Fig. 5.13: X-ray diffraction peaks of aluminum alloy.

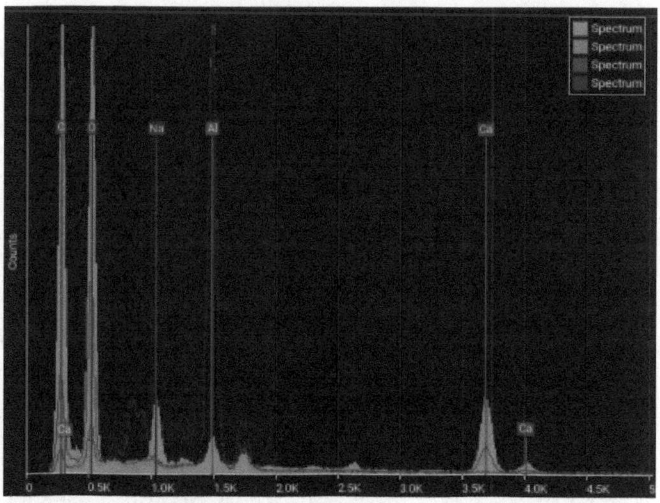

Fig. 5.14: X-ray diffraction peaks of aluminum alloy and CNT composite.

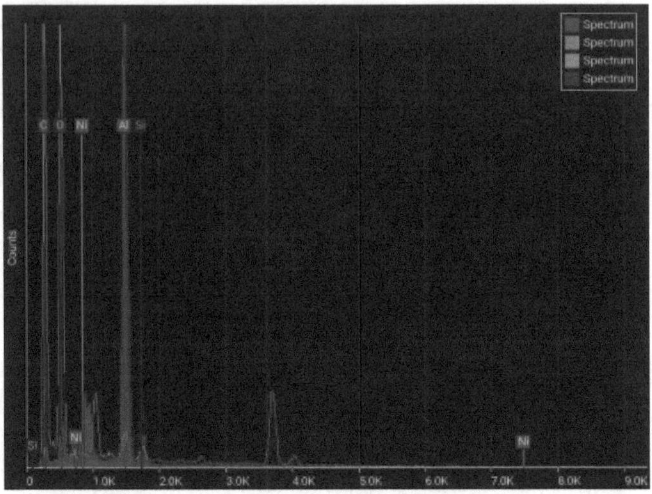

Fig. 5.15: X-ray diffraction peaks of aluminum alloy, CNT and silicon carbide composite.

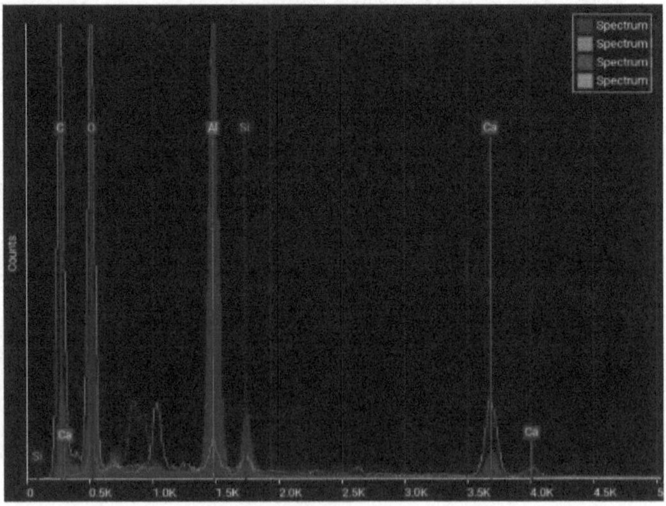

Fig. 5.16: X-ray diffraction peaks of aluminum alloy, CNT, silicon carbide and alumina composite.

Presence of carbon could be due to the conversion of aluminum atoms at higher temperature, since the working temperature was higher than the melting point of aluminum. Similarly, presence of oxygen is observed to be high, which could possibly be due to the formation of oxides of aluminum. Presence of calcium and nickel was noted to be almost constant in all the samples which could be attributed to the degradation of carbon nanotube, silicon carbide and alumina powder.

5.3 Physical Properties

Physical characteristics of the newly developed aluminum metal matrix composites have been tested and recorded for necessary assessment.

5.3.1 Density and Porosity

Theoretical density, experimental density, and percent porosity of composites are presented in Table 5.1. It can be inferred that the measured values for aluminum alloy are closer to the theoretical values, as the reinforced content increases the deviation in the theoretical and measured values increases. It indicates that, as the reinforcement wt.% increases, the percent porosity increases. It is because of the cavity formed in the composites.

Table 5.1: Density and percentage porosity of hybrid particle reinforced aluminum metal matrix composites

Serial no.	Composition	Sample	Theoretical density (g/cc)	Experimental density (g/cc)	Percentage porosity (%)
1	Al-alloy	Specimen 1	2.650	2.627	0.868
2	Al-alloy+1 % CNT	Specimen 2	2.638	2.588	1.875
3	Al-alloy+1% CNT+ 2.5% SiC	Specimen 3	2.652	2.592	2.263
4	Al-alloy+1% CNT+ 2.5% SiC+ 2.5% Al_2O_3	Specimen 4	2.684	2.605	2.937

Again, small variation of densities and in experimental and measured values is probably because some of the reinforcements may shift into the slag. The density of the carbon nanotube was 1.4g/cc, silicon carbide 3.21g/cc, and alumina 3.96g/cc against the measured density of 2.63g/cc for the base aluminum alloy. The density of the composites reinforced with mono-particle (carbon nanotube) decreased by 1.47%. Virat et al. reported that the density of composites reinforced with carbon nanotubes decreased. With bi-particle (carbon nanotube and silicon carbide) reinforced composite the density is decreased by 1.32% compared to base aluminum alloy but is increased by 0.15% compared to mono-particle reinforced composite. For the case of tri-particle (carbon nanotube, silicon carbide, and alumina) reinforced composite the density is decreased by 0.83% compared to base alloy but is increased by 0.65% and 0.50% compared to mono-particle reinforced and bi-particle reinforced composites respectively. This is due to higher and lower value densities and the wt.% addition of the reinforcements.

The porosity increased up to 238.36% for tri-particle reinforced composite compared to the base aluminum alloy. It is because the greater density of silicon carbide and alumina reinforcements; air and gas trapped with constituents' fragments during mixing, cavity shrinkage during solidification, or molecular oxygen evolution could all contribute to this increase in permeability or porosity.

From Fig. 5.17, it can be shown that, as the reinforcement content increases, the density of the composite specimens reduced. This is due to higher and lower value densities and the wt.% addition of the reinforcements. Fig. 5.18 indicates that the reinforcement percentage content increases, the percent porosity increased.

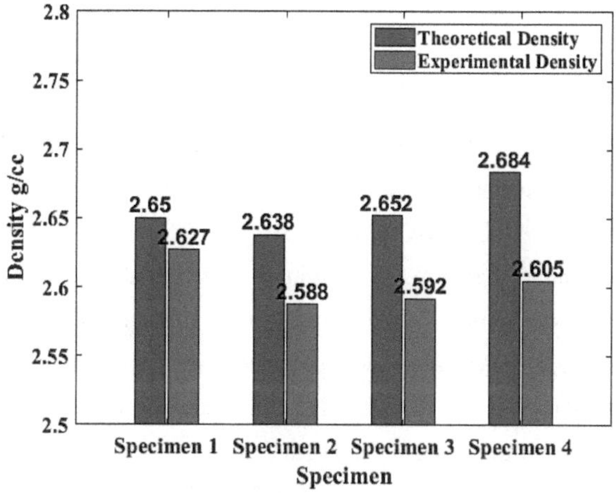

Fig. 5.17: Theoretical density and experimental density of aluminium metal matrix composites.

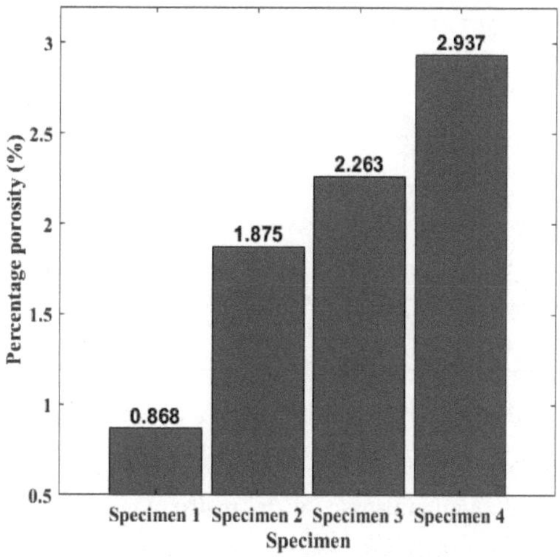

Fig. 5.18: Percentage porosity of aluminium metal matrix composites.

5.4 Mechanical Properties

Mechanical and Morphological characteristics of the newly developed aluminum metal matrix composites have been tested and recorded for necessary assessment.

5.4.1 Tensile Strength

The addition of carbon nanotube, silicon carbide, and alumina increases the UTS and yield strength of the composites. UTS increased to a maximum of 142.857% and yield strength increased by about 143.75% compared to base aluminum alloy while adding bi-particle reinforcements (1% carbon nanotube and 2.5% silicon carbide) in the aluminum alloy. An increase in tensile strength is generally attributed to the strong interfacial bonding between the aluminum matrix and reinforcements. For mono-particle reinforced composite, the ultimate tensile strength was 121.428% and the yield strength of 159.375 increased compared to the base alloy. Similar kinds of increases in tensile strength while adding carbon nanotube particles were reported by researchers. The effect of the ultimate tensile strength (UTS) and yield strength of the aluminium-based metal matrix composites is shown in Table 5.2.

Table 5.2: Tensile strength variation of each composition

Serial no.	Composition	Sample	Tensile strength (MPa)	Yield strength (MPa)	Elongation (%)
1	Al-alloy	Specimen 1	42	32	1.6
2	Al-alloy+1 % CNT	Specimen 2	93	83	5.7
3	Al-alloy+1% CNT+ 2.5% SiC	Specimen 3	102	78	5.8
4	Al-alloy+1% CNT+ 2.5% SiC+ 2.5% Al_2O_3	Specimen 4	96	62	5.2

However, the addition of alumina for tri-particle reinforced metal matrix composites decreases the tensile strength by 6.25% and yield strength by 20.31% compared to bi-particle reinforced aluminum-based metal matrix composite. When compared to base aluminum alloy, for tri-particle reinforced composite the UTS increased by about 128.57% and yield strength increased by about 93.75%.

The addition of mono-particle and bi-particle reinforcements in the aluminum alloy increases the elongation of the composites compared to base alloy, but for tri- particle reinforcements, the elongation decreased by about 14.78% compared to bi-particulate reinforced composite. The presence of hard alumina decreases the ductility of composites, thereby encouraging plastic deformation and hence a decrease in elongation in the composite.

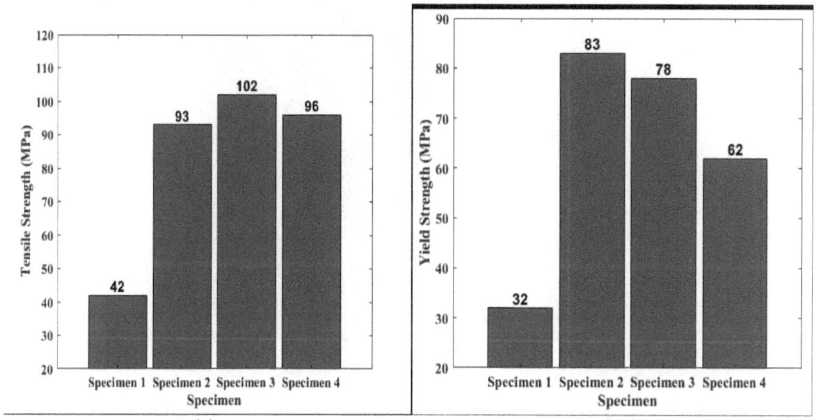

(a) (b)

Fig. 5.19: a) Ultimate tensile strength (UTS), b) Yield tensile strength of Aluminium metal matrix composites.

The addition of carbon nanotube, silicon carbide, and alumina increases the UTS and yield strength of the composites, shown in Fig. 5.19. Multidirectional grain refinement at the matrix/reinforcement interface induces the strengthening effect of the composites, thereby increasing the tensile strength of the composites (Chen et al., 2017).

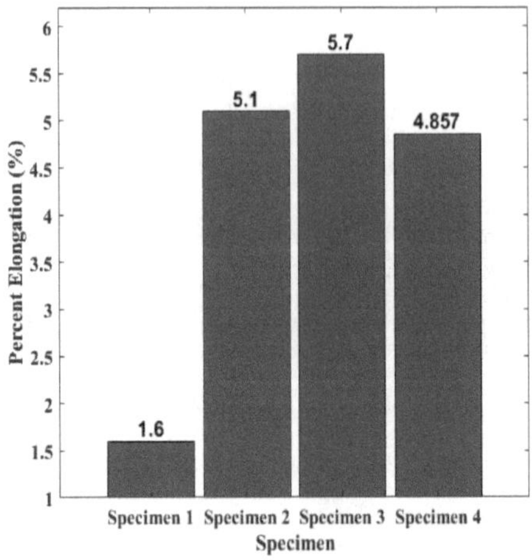

Fig. 5.20: Percentage elongation of aluminium metal matrix composites.

The addition of mono-particle and bi-particle reinforcements in the aluminum alloy increases the elongation of the composites compared to base alloy as per Fig. 5.20, but for tri-particle reinforcements, the elongation decreased. The presence of hard alumina decreases the ductility of composites, thereby encouraging plastic deformation and hence a decrease in elongation in the composite.

5.4.2 Flexural Strength

The effect of the flexural strength of the aluminum-based metal matrix composites is shown in Table 5.3.

Table 5.3: Flexural strength variation of each composition

Serial no.	Composition	Sample	Flexural strength (kN)
1	Al-alloy	Specimen 1	1.21
2	Al-alloy+1 % CNT	Specimen 2	3.547
3	Al-alloy+1% CNT+ 2.5% SiC	Specimen 3	3.386
4	Al-alloy+1% CNT+ 2.5% SiC+ 2.5% Al_2O_3	Specimen 4	3.42

The results reveal that the inclusion of particles increased the flexural strength by about 193.14% for mono-particle reinforced, 179.63% for bi-particle reinforced, and

182.64% for tri-particle reinforced metal matrix composite compared to the base alloy. It indicates, due to the inclusion of reinforcement, the inter-spatial distance between composite elements is reduced which results in increasing the load distribution capacity of the composite.

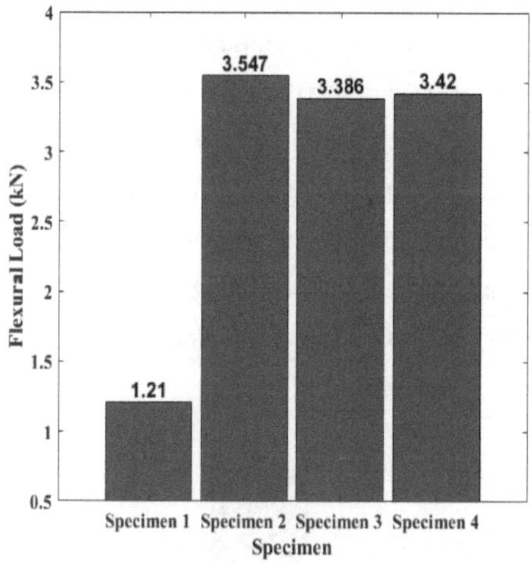

Fig. 5.21: Flexural strength of aluminium metal matrix composites.

Fig. 5.21 reveal that the inclusion of particles increased the flexural strength.

5.4.3 Impact Strength

The influence on impact strength while adding multiple particulate reinforcements in aluminum metal matrix composite is shown in Table 5.4. The Charpy impact test was used to determine impact strength in this case. The toughness of the composites as measured by the Charpy impact tester shows that as the reinforcing content increases, the toughness of the composite falls.

Table 5.4: Impact strength variation of each composition

Serial no.	Composition	Sample	Impact strength (J)
1	Al-alloy	Specimen 1	1
2	Al-alloy+1 % CNT	Specimen 2	2

Table 5.4 (Continued)			
Serial no.	Composition	Sample	Impact strength (J)
3	Al-alloy+1% CNT+ 2.5% SiC	Specimen 3	1.6
4	Al-alloy+1% CNT+ 2.5% SiC+ 2.5% Al_2O_3	Specimen 4	1.45

The results show, with the addition of carbon nanotubes the impact strength increases by 100% comparing the base alloy as carbon nanotubes tend to increase the ductility of the composite for its intrinsic property. It increases the inter-spatial bonding between the elements in the composites. For bi-particle and tri-particle reinforced metal matrix composite the impact strength was found respectively 20% and 27.5% less than compared to mono-particle reinforced metal matrix composite. With the increasing amount of silicon carbide and alumina, composite properties tend to transom from ductile to brittle which causes a reduction in impact strength.

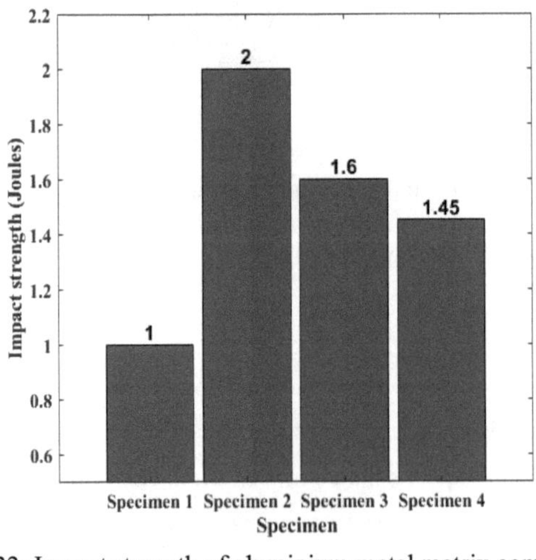

Fig. 5.22: Impact strength of aluminium metal matrix composites.

The influence on impact strength while adding multiple particulate reinforcements in aluminum metal matrix composite is shown in Fig. 5.9. As the reinforcing content increases, the toughness of the composite falls. With the inclusion of silicon carbide and alumina, composite properties tend to transform from ductile to brittle which causes a reduction in impact strength.

5.4.4 Hardness

A rectangular specimen of 55x55x10 mm^3 was cast for hardness testing and afterward machined into a standard specimen. A finely polished plate surface was used for the Brinell hardness test. The work piece was placed beneath a diamond indenter for 5 sets of impressions, and the Brinell hardness number (BHN) was calculated on average. The hardness of aluminum metal matrix composites is shown in Table 5.5. It depicts the hardness result achieved by the Brinell hardness test. It demonstrates that the hardness of aluminum metal matrix composites increases as the reinforcement added.

Table 5.5: Hardness variation of each composition

Serial no.	Composition	Sample	Hardness in (BHN)
1	Al-alloy	Specimen 1	126.420
2	Al-alloy+1 % CNT	Specimen 2	136.975
3	Al-alloy+1% CNT+ 2.5% SiC	Specimen 3	146.306
4	Al-alloy+1% CNT+ 2.5% SiC+ 2.5% Al_2O_3	Specimen 4	135.437

Fig. 5.23 demonstrates that the hardness of aluminum metal matrix composites increases with the inclusion of reinforcement. The enhanced hardness and strength are due to the strengthening effect of the dispersed reinforcement particles and of the refined microstructure. The nanoparticles can interact with dislocations, hampering their movement and leading to dislocations bowing around the particles, is considered by Orowan strengthening effect (Zhang et al., 2006; Zhang et al., 2008).

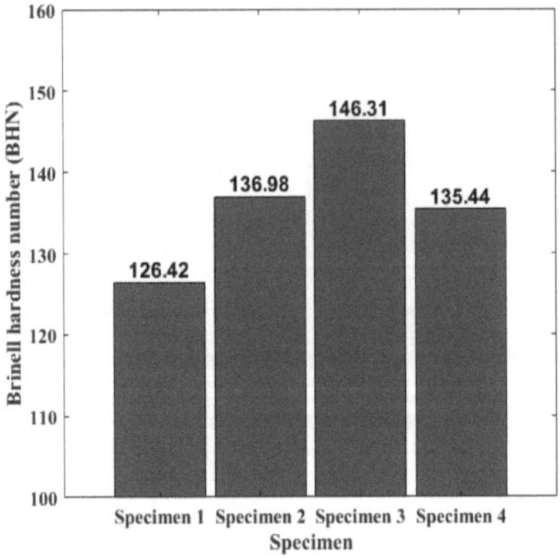

Fig. 5.23: Hardness of aluminium metal matrix composites.

5.5 Machinability test

The evaluation of drilled hole quality was done by calculating the roundness deviation, delamination factor and taper angle. 4 different specimens were experimented with different particle wt.% reinforcement as mentioned and the following results have been observed throughout the experiment.

5.5.1 Roundness Deviation

From the experiment, some variations in the diameter have been observed at the top and bottom surface of the specimens. A digital slide calipers was used to measure the hole roundness of the specimens with 0.1mm least count. Table 5.6 to Table 5.9, shows the influence of the reinforcement weight percentage on inlet (top) diameter and outlet (bottom) diameter of the holes respectively. The roundness deviation value was affected by changing the operating conditions such as the feed rate and cutting velocity. The cutting speed was a very strong influence on the surface quality of the drilled hole.

The results of the experiment indicate that using a drill bit of 8mm, the inlet diameter of the holes was varied with the variation of rotational speed and feed rate. At low rotational speed it showed, larger diameter in the inlet holes than at higher rotational speed. But in case of outlet diameter, they were affected and decreased to below 8mm sometimes. This is due to the lack of lubrication effect. The damages related to temperature occurred due to friction between the drill and the wall of the hole and caused roundness deviation. With hybrid particulate reinforcements, as thermal conductivity of the reinforcement particles is higher than the base aluminum, this is why in high particulate reinforcement content specimen, the diameter deviation was less compared to other specimens.

Table 5.6: Diameter deviation in the top and bottom holes of Aluminum alloy specimen

Specimen 1- Al-alloy					
RPM	Feed rate, mm/rev	In diameter (Top), mm	Out diameter (Bottom), (mm)	Taper angle	Delamination
	0.1	8.5	7.9	0.0299	1.2613
400	0.15	8.5	8	0.02499	1.2611
	0.2	8.45	8	0.0225	1.2537
	0.1	8.2	8	0.0099	1.2166
1140	0.15	8.2	8	0.0099	1.2166
	0.2	8.1	8	0.0049	1.2018
	0.1	8.1	7.9	0.0099	1.2018
2260	0.15	8.1	7.9	0.0099	1.2018
	0.2	8.1	8	0.0049	1.2018

Table 5.7: Diameter deviation in the top and bottom holes of Aluminum alloy + CNT specimen

Specimen 2- Al-alloy + CNT					
RPM	Feed rate, mm/rev	In diameter (Top), mm	Out diameter (Bottom), (mm)	Taper angle	Delamination
400	0.1	8.3	7.9	0.0199	1.2315
	0.15	8.3	8	0.0149	1.2315
	0.2	8.2	7.85	0.0175	1.2166
1140	0.1	8.1	7.9	0.0099	1.2018
	0.15	8.1	7.9	0.0099	1.2018
	0.2	8.2	8	0.0099	1.2166
2260	0.1	8.2	8.1	0.0049	1.2166
	0.15	8.1	8	0.0049	1.2018
	0.2	8.15	8	0.0075	1.2092

Table 5.8: Diameter deviation in the top and bottom holes of Aluminum alloy + CNT + SiC specimen

Specimen 3- Al-alloy + CNT +SiC					
RPM	Feed rate, mm/rev	In diameter (Top), mm	Out diameter (Bottom), (mm)	Taper angle	Delamination
400	0.1	8.5	7.9	0.0299	1.2611
	0.15	8.2	8	0.0099	1.2166
	0.2	8.3	8.1	0.0099	1.2315
1140	0.1	8.1	7.9	0.0099	1.2018
	0.15	8.2	7.9	0.0149	1.2166
	0.2	8.1	7.9	0.0099	1.2018
2260	0.1	8.1	8	0.0049	1.2018
	0.15	8.1	7.9	0.0099	1.2018
	0.2	8.2	8	0.0099	1.2166

Table 5.9: Diameter deviation in the top and bottom holes of Al + CNT + SiC + Al_2O_3 specimen

Specimen 3- Al + CNT + SiC + Al_2O_3					
RPM	Feed rate, mm/rev	In diameter (Top), mm	Out diameter (Bottom), (mm)	Taper angle	Delamination
	0.1	8.3	8	0.0149	1.2315
400	0.15	8.2	7.9	0.0149	1.2167
	0.2	8.1	8	0.0049	1.2018
	0.1	8.2	8	0.0099	1.2167
1140	0.15	8.2	7.9	0.0149	1.2167
	0.2	8.1	7.9	0.0099	1.2018
	0.1	8.1	8	0.0049	1.2018
2260	0.15	8.15	8	0.0075	1.2092
	0.2	8	7.9	0.0049	1.1869

From the experiment, some variations in the diameter have been observed at the top and bottom surface of the specimens and shown in Fig. 5.24 to Fig. 5.27.

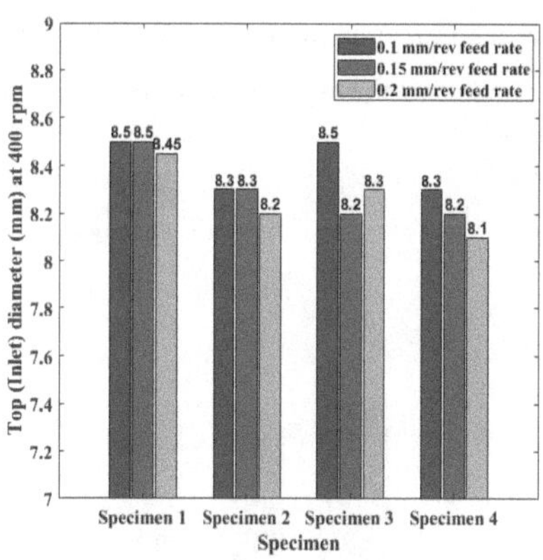

(a)

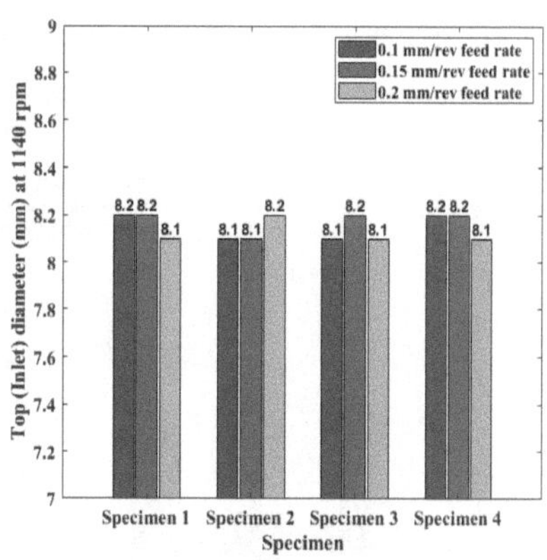

(b)

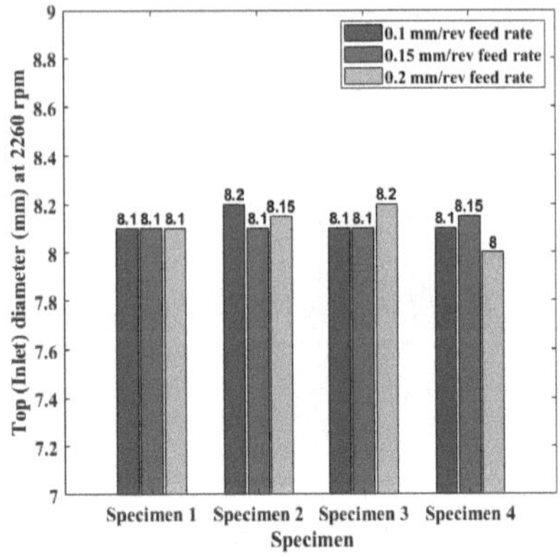

(c)

Fig. 5.24: Effect of rotational speed and feed rate in Top (Inlet) diameter (a) at 400 rpm, (b) at 1140 rpm, (c) at 2260 rpm.

At low rotational speed, it showed a larger diameter in the inlet holes than at higher rotational speed, shown in Fig. 5.24. But in the case of outlet diameter shown in Fig. 5.25, they were affected and decreased to below 8mm sometimes. This is due to the lack of lubrication effect. The damages related to temperature occurred due to friction between the drill and the wall of the hole and caused roundness deviation.

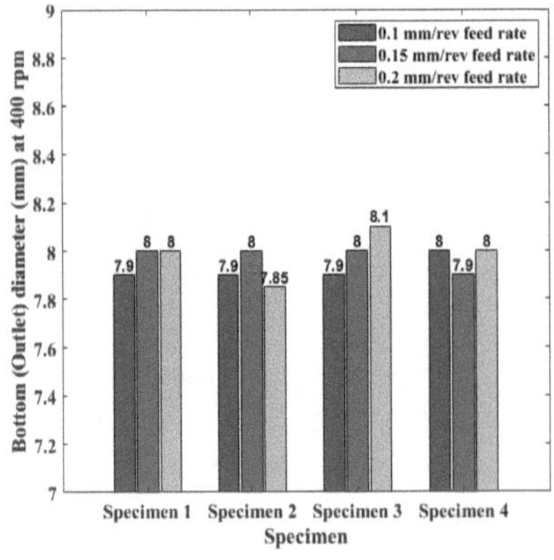

(a)

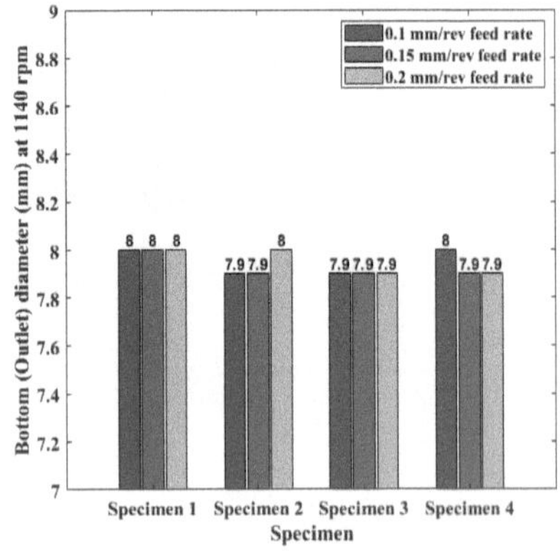

(b)

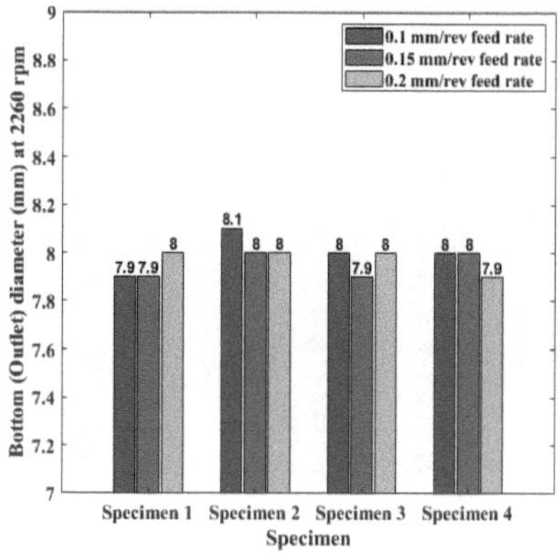

(c)

Fig. 5.25: Effect of rotational speed and feed rate in Bottom (Outlet) diameter (a) at 400 rpm, (b) at 1140 rpm, (c) at 2260 rpm.

The cutting speed was a very strong influence on the surface quality of the drilled hole.

5.5.2 Delamination

As various spindle speeds in this experiment, drilling induced damage was observed in terms of both quantity and quality. For the calculation of quantitative term delamination factor (F_d), the given formula was used.

$$F_d = D_{max}/D_{hole} \tag{5.1}$$

Where, F_d is the conventional delamination factor, D_{hole} is the hole exit diameter and D_{max} is the damaged zone diameter. Table 5.6 to Table 5.9 represents the calculation of the delamination factor for four different specimens.

It is clear that maximum metal pull-out is found at the specimen-1 (Aluminum–alloy). On the other hand, specimen-4 (Aluminum metal matrix composite reinforced with SiC, Al_2O_3 and CNT) shows lower delamination factor.

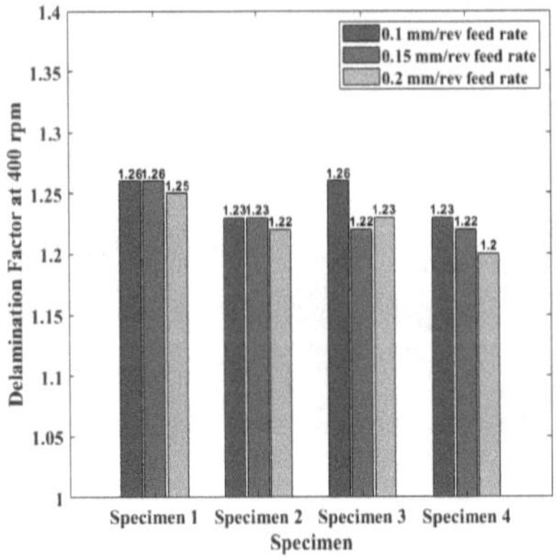

(a)

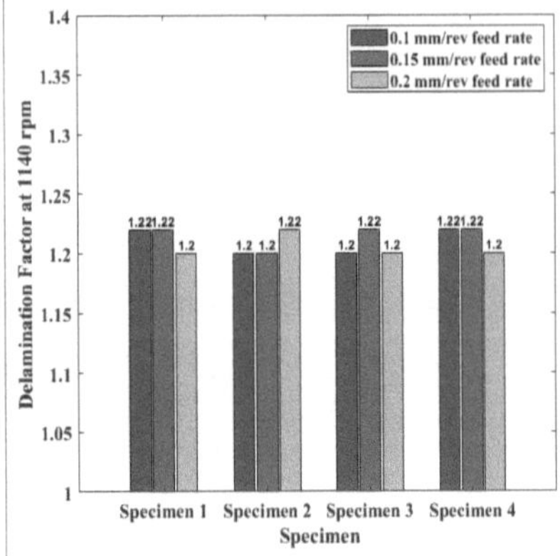

(b)

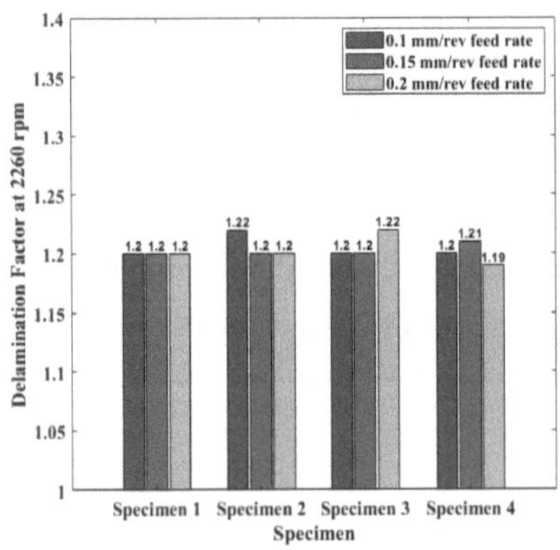

(c)

Fig. 5.26: Delamination for different metal matrix composites (a) at 400 rpm, (b) at 1140 rpm, (c) at 2260 rpm.

For composite lower delamination factor is found and shown in Fig. 5.26, which indicates lower metal pull-out with minimum surface roughness. This is due to having higher weight percentage of particle reinforcement contents, which allows its matrix to distributed the stress generated while drilling more effectively than the three other specimens. As a result, less metal pull-out is observed which leads to minimum surface roughness in specimen-4. Furthermore, as the cutting speed increases, the delamination factor decreases for particle reinforced composites and the best quality surface of the drilled hole obtained.

5.5.3 Taper Angle

The taper angle is dependent on the focal position regarding the work piece surface, focus depth, drilled hole depth, duration of the pulse etc. In this case, the taper angle is produced due to the dissimilarity between the inlet and outlet diameter of the HAMMCs plate. The following formula has been used to measure the taper angle.

$$\delta = tan^{-1}((D_{in} - D_{out})/2L) \qquad (5.2)$$

Where, D_{in} indicates the hole's entrance diameter, D_{out} indicates the hole's exit diameter and L indicates specimen's thickness.

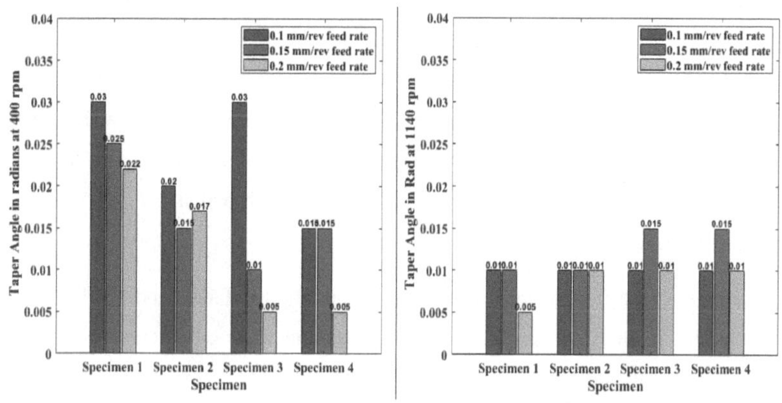

(a) (b)

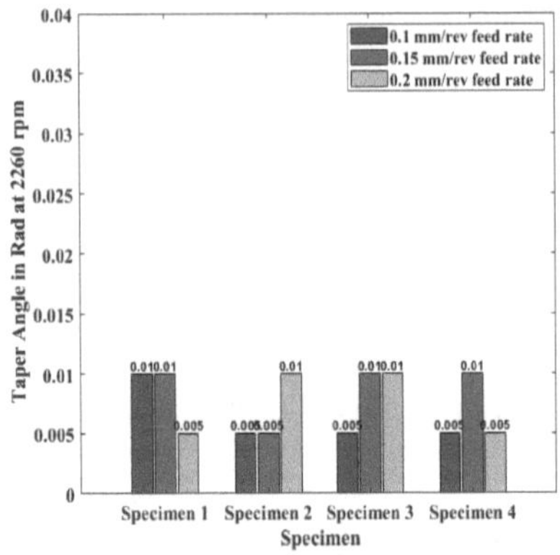

(c)

Fig. 5.27: Taper angle in radians for different metal matrix composites (a) at 400 rpm, (b) at 1140 rpm, (c) at 2260 rpm.

From Fig. 5.27, taper angle variation was observed under dry condition. From this observation, the average taper values were noticed to be positive, which shows, the inlet (top) diameters of the holes were found larger than the outer (bottom) diameters. The radial vibration is greater at the entry than the exit which caused a positive value of the taper angle.

After assessing the roundness deviation, delamination factor and finally calculating taper angle values, it was found that, the specimen having higher wt.% of particulate reinforcements has the better hole quality than the specimens having single or double particulate reinforcements wt.%. That means the specimen containing silicon oxide, alumina and carbon nano-tube is more suitable to perform drilling operation at high cutting speed and feed rate value. Better quality holes are the ultimate result of better machinability which is mandatory to use metal matrix composite in industrial and constructional sectors.

5.6 Machining Parameters Optimization

Choice of the process parameters that would produce an excellent surface quality with ensuring low tooling cost is the main challenges for today's manufacturers. Thus, there is a requirement of selection of the best combination of input parameters to generate the desired responses.

Machining parameters have a major effect on the quantity and quality of the produced part and cost of production. The selected machining parameters should yield desired need while utilizing the machining resources such as machine tool and cutting tool to the fullest extent with considering the associated constraints on these resources. Hence, the optimization of the parameters is essential. The aim of the optimization is to improve an existing process that meets the given requirements and satisfies all the constraints placed on it. As one of the objectives of this study is to get optimum values of cutting speed and feed while minimizing roundness deviation, delamination and taper angle.

5.6.1 Optimization of Drilling Parameters

The grey relational analysis based on grey system theory can be used for solving the complicated interrelationships among the multi responses. A gray relational grade is obtained to evaluate the multiple responses. As a result, optimization of the multiple responses can be converted into optimization of a single relational grade. In short, there is an ample scope of applying the proposed methodology of grey relational analysis and Taguchi method with the multiple responses for the optimization of drilling parameters.

5.6.2 Optimization Steps Using Grey Relational Analysis

The factors and their levels considered in this study are shown in Table 5.10. Experiments are conducted with two factors each at three levels and hence a three level orthogonal array (OA) is chosen.

Table 5.10 Factors and Levels for Grey Relational Analysis

Parameters	Unit	Level		
		1	2	3
Cutting Speed (V)	m/min	400	1140	2260
Feed (F)	Mm/rev	0.1	0.15	0.2

The following tables shows the matrix for developed composites.

Table 5.11: Orthogonal array with factors and responses for Al-alloy

Specimen 1- Al-alloy					
RPM	Feed rate, mm/rev	In diameter (Top), mm	Out diameter (Bottom), (mm)	Taper angle	Delamination
400	0.1	8.5	7.9	0.0299	1.2613
	0.15	8.5	8	0.02499	1.2611
	0.2	8.45	8	0.0225	1.2537
1140	0.1	8.2	8	0.0099	1.2166
	0.15	8.2	8	0.0099	1.2166
	0.2	8.1	8	0.0049	1.2018
2260	0.1	8.1	7.9	0.0099	1.2018
	0.15	8.1	7.9	0.0099	1.2018
	0.2	8.1	8	0.0049	1.2018

Table 5.12: Orthogonal array with factors and responses for Al-alloy and CNT composites

Specimen 2- Al-alloy + CNT					
RPM	Feed rate, mm/rev	In diameter (Top), mm	Out diameter (Bottom), (mm)	Taper angle	Delamination
400	0.1	8.3	7.9	0.0199	1.2315
	0.15	8.3	8	0.0149	1.2315
	0.2	8.2	7.85	0.0175	1.2166
1140	0.1	8.1	7.9	0.0099	1.2018
	0.15	8.1	7.9	0.0099	1.2018
	0.2	8.2	8	0.0099	1.2166
2260	0.1	8.2	8.1	0.0049	1.2166
	0.15	8.1	8	0.0049	1.2018
	0.2	8.15	8	0.0075	1.2092

Table 5.13: Orthogonal array with factors and responses for Al-alloy, SiC and CNT composites

Specimen 3- Al-alloy + CNT +SiC					
RPM	Feed rate, mm/rev	In diameter (Top), mm	Out diameter (Bottom), (mm)	Taper angle	Delamination
400	0.1	8.5	7.9	0.0299	1.2611
	0.15	8.2	8	0.0099	1.2166
	0.2	8.3	8.1	0.0099	1.2315
1140	0.1	8.1	7.9	0.0099	1.2018
	0.15	8.2	7.9	0.0149	1.2018
	0.2	8.1	7.9	0.0099	1.2018
2260	0.1	8.1	8	0.0049	1.2018
	0.15	8.1	7.9	0.0099	1.2018
	0.2	8.2	8	0.0099	1.2166

Table 5.14: Orthogonal array with factors and responses for Al-alloy, SiC, alumina and CNT composites

Specimen 4- Al + CNT + SiC + Al_2O_3					
RPM	Feed rate, mm/rev	In diameter (Top), mm	Out diameter (Bottom), (mm)	Taper angle	Delamination
400	0.1	8.3	8	0.0149	1.2315
	0.15	8.2	7.9	0.0149	1.2167
	0.2	8.1	8	0.0049	1.2018
1140	0.1	8.2	8	0.0099	1.2167
	0.15	8.2	7.9	0.0149	1.2166
	0.2	8.1	7.9	0.0099	1.2018
2260	0.1	8.1	8	0.0049	1.2018
	0.15	8.15	8	0.0075	1.2092
	0.2	8	7.9	0.0049	1.1869

Based on the calculation on appendix F, considering maximization of grade values from Table F.18 to Table F.21, we get for which composite the optimal parameter conditions of cutting speed and feed rate can be obtained.

Table 5.15: Grey relational grade for aluminum alloy, silicon carbide, alumina and CNT composites

Specimens	RPM	Feed rate, mm/rev	Grey relation grade	Ranking
Al-alloy	2260	0.1	0.8909	1
		0.15	0.8909	1
Al –alloy + CNT	2260	0.1	0.3736	1
Al –alloy + CNT + Silicon carbide	2260	0.1	0.7615	1
Al –alloy + CNT + Silicon carbide + alumina	2260	0.2	1.0365	1

From table 5.15, we observed that, the optimal parameter condition we can be achieved for unreinforced aluminum alloy in two cases, for 2260rpm at 0.1 mm/rev and 0.15 mm/rev feed rate. For the case of mono-particle reinforced aluminum matrix composite, the optimum parameter is 2260rpm at 0.1 mm/rev feed rate and for bi-particle reinforced aluminum matrix composite, the optimum parameter is 2260rpm at 0.1 mm/rev feed rate. Whereas, for tri-particle reinforced aluminum matrix composite the optimum parameters were 2260rpm and 0.2 mm/rev feed rate. So, it can be concluded that, with tri-particle reinforced composite we obtained better machining quality in products with minimal machining time required.

The optimum case also could be obtain using the Taguchi method. Using the data presented in Table 5.16 to 5.23, it is also possible to find the best case for machining for each samples.

Table 5.16: Response table for signal to noise ratio for aluminum alloy

Specimen 1- Al –alloy Smaller the better Level	RPM	Feed rate
1	-18.016	-10.294
2	-12.5	-11.021
3	-3.181	-12.38
Delta	14.835	2.088
Rank	1	2

Table 5.17: Response table for means for aluminum alloy

Specimen 1- Al –alloy Smaller the better Level	RPM	Feed rate
1	8	4.333
2	4.333	5
3	1.667	4.667
Delta	6.333	0.667
Rank	1	2

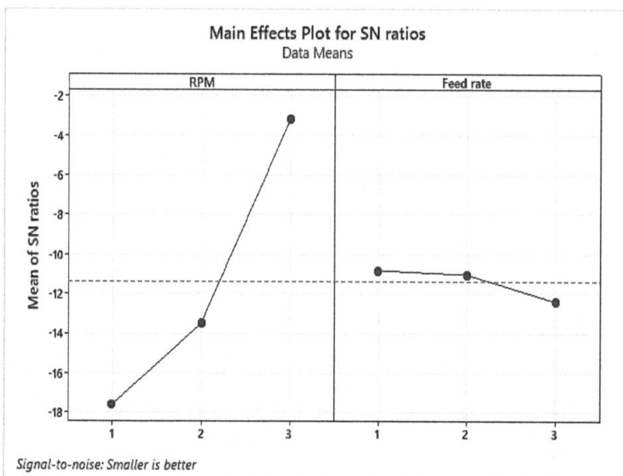

Fig. 5.28: Factor effects plot for SN ratio for aluminum alloy.

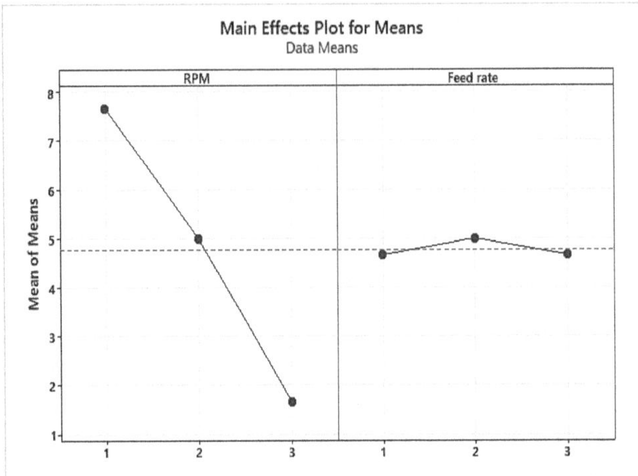

Fig. 5.29: Factor effects plot for means for aluminum alloy.

From Table 5.16 and Table 5.17 and Fig 5.28 and Fig. 5.29, considering smaller is better we found the optimal parameters for unreinforced aluminum alloy is RPM at 3rd level and feed rate at 1st level. Also, RPM is the dominant factor for aluminum alloy.

Table 5.18: Response table for signal to noise ratio for aluminum alloy and CNT composite

Specimen 2- Al –alloy + CNT		
Smaller the better		
Level	RPM	Feed rate
1	-15.035	-6.667
2	-10.034	-13.556
3	-10.822	-15.662
Delta	5	9
Rank	2	1

Table 5.19: Response table for means for aluminum alloy and CNT composite

Specimen 2- Al –alloy + CNT		
Smaller the better		
Level	RPM	Feed rate
1	6	2.667

Specimen 2- Al –alloy + CNT		
Smaller the better		
Level	RPM	Feed rate
2	4	5.667
3	4.667	6.33
Delta	2	3.667
Rank	2	1

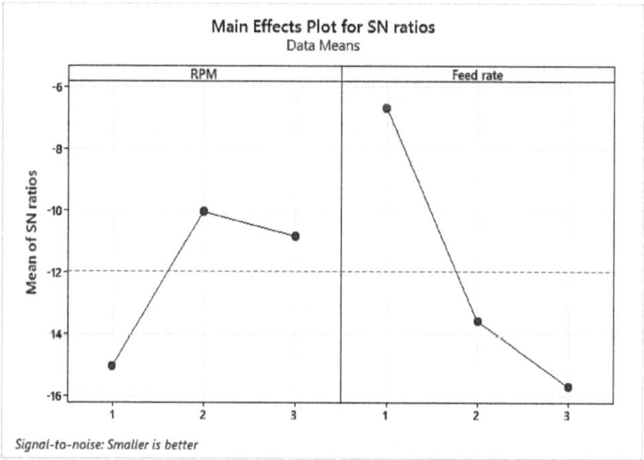

Figure 5.30: Factor effects plot for SN ratio for aluminum alloy and CNT composite.

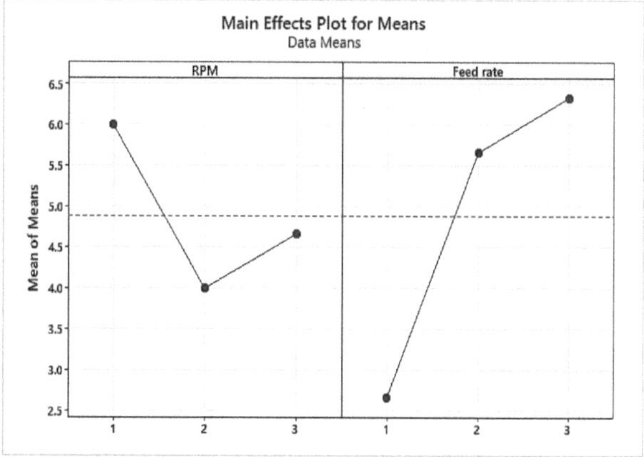

Figure 5.31: Factor effects plot for mean for aluminum alloy and CNT composite.

From Table 5.18 and Table 5.19 and Fig 5.30 and Fig. 5.31, considering smaller is better we found the optimal parameters for mono-reinforced aluminum matrix composite is RPM at 2nd level and feed rate at 1st level. Also, feed rate is the dominant factor for the composite.

Table 5.20: Response table for signal to noise ratio for aluminum alloy, CNT and silicon carbide composite

Specimen 3- Al –alloy + CNT + silicon carbide		
Smaller the better		
Level	RPM	Feed rate
1	-12.382	-11.0210
2	-15.34	-12.687
3	-6.667	-10.68
Delta	8.674	2.007
Rank	1	2

Table 5.21: Response table for means for aluminum alloy, CNT and silicon carbide composite

Specimen 3- Al –alloy + CNT + silicon carbide		
Smaller the better		
Level	RPM	Feed rate
1	5	5
2	6	5
3	2.667	3.667
Delta	3.33	1.33
Rank	1	2

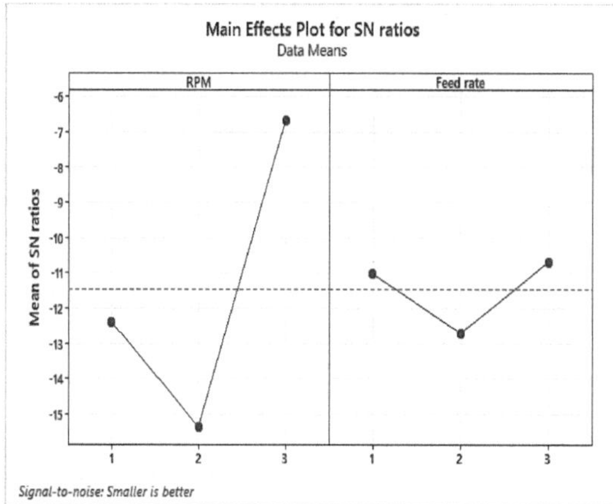

Fig. 5.32: Factor effects plot for SN ratio for aluminum alloy, CNT and silicon carbide composite.

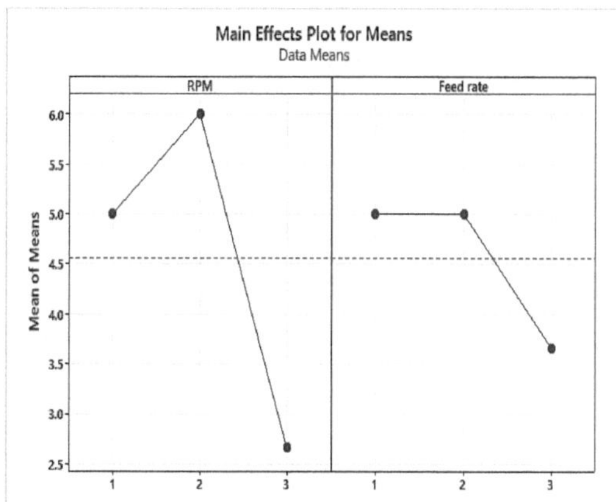

Fig. 5.33: Factor effects plot for mean for aluminum alloy, CNT and silicon carbide composite.

From Table 5.20 and Table 5.21 and Fig 5.32 and Fig. 5.33, considering smaller is better we found the optimal parameters for bi-particle reinforced aluminum matrix composite is RPM at 3rd level and feed rate at 3rd level. Also, RPM is the dominant factor for the composite.

Table 5.22: Response table for signal to noise ratio for aluminum alloy, CNT, silicon carbide and alumina composite

Specimen 4- Al –alloy + CNT + silicon carbide + alumina		
Smaller the better		
Level	RPM	Feed rate
1	-13.556	-16.009
2	-13.275	-14.002
3	-9.201	-6.021
Delta	4.355	9.989
Rank	2	1

Table 5.23: Response table for means for aluminum alloy, CNT, silicon carbide and alumina composite

Specimen 4- Al –alloy + CNT + silicon carbide + alumina		
Smaller the better		
Level	RPM	Feed rate
1	5.333	6.667
2	5.333	5.333
3	3.667	2.333
Delta	1.667	4.333
Rank	2	1

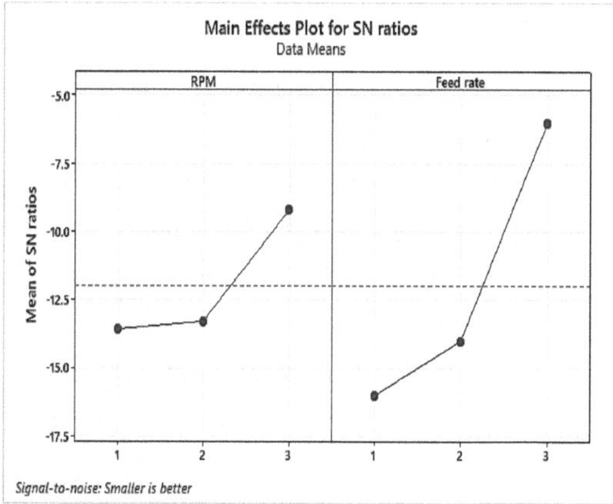

Fig. 5.34: Factor effects plot for SN ratio for aluminum alloy, CNT, silicon carbide and alumina composite.

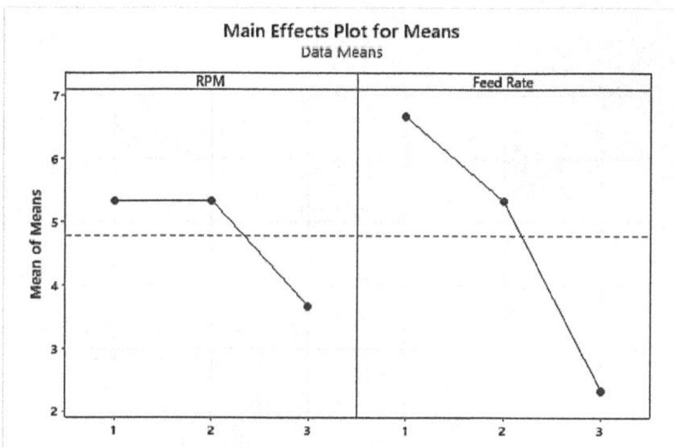

Fig. 5.35: Factor effects plot for mean for aluminum alloy, CNT, silicon carbide and alumina composite.

From Table 5.22 and Table 5.23 and Fig 5.34 and Fig. 5.35, considering smaller is better we found the optimal parameters for tri-particle reinforced aluminum matrix composite is RPM at 3rd level and feed rate at 3rd level. Also, feed rate is the dominant factor for the composite.

5.7 Chip Morphology

During the drilling operation of aluminum alloys, usually small and segmented chips are desirable for the high-quality hole and increased tool life. Chips quality for aluminum alloy and the composites are shown in Fig. 5.36 to Fig. 5.39. From the formation of the chips, it can be found that for aluminum alloy the chips have blunter edges and size of the chips is bigger compared to the chips formed in the composites. For composite, as the reinforcement content increases, the chips size getting shorter and edge were sharper compared to aluminum alloy. This indicates the delamination reduction in machining process and improvement in machining quality.

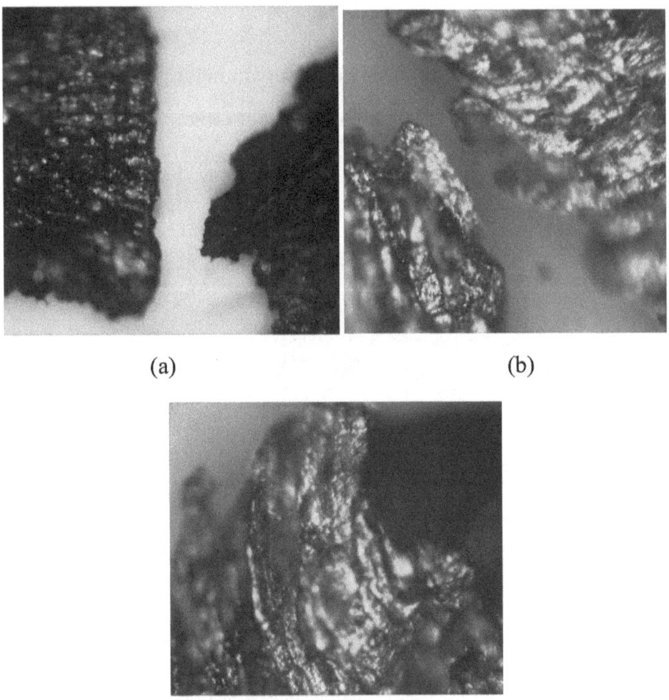

(a) (b)

(c)

Fig. 5.36: Chips for aluminum alloy at, a) 80x magnification, b) 160x magnification, c) 320x magnification.

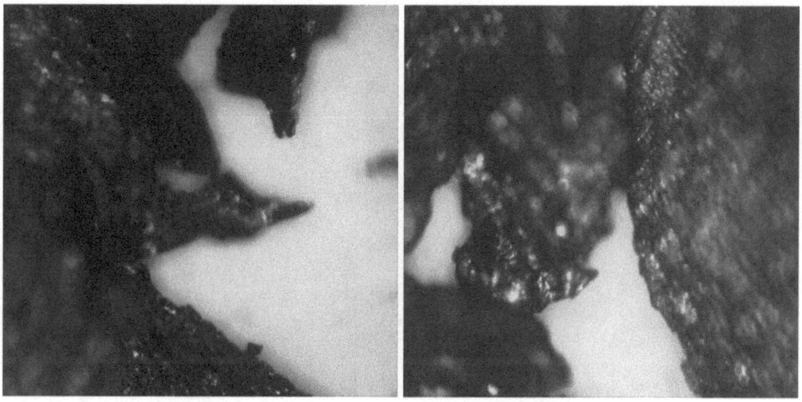

(a) (b)

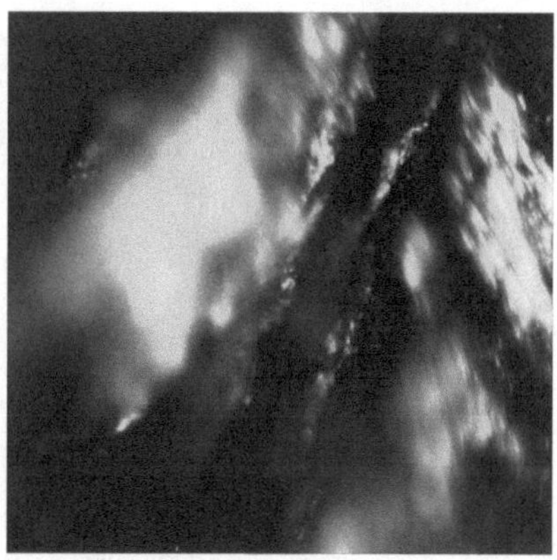

(c)

Fig. 5.37: Chips for aluminum alloy and CNT composite at, a) 80x magnification, b) 160x magnification, c) 320x magnification.

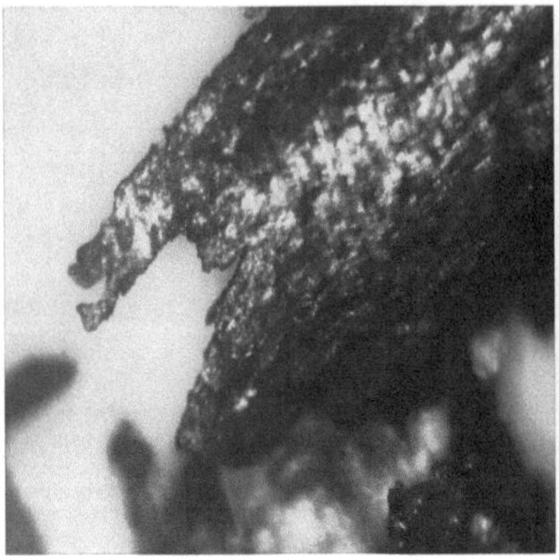

(a)

(b)

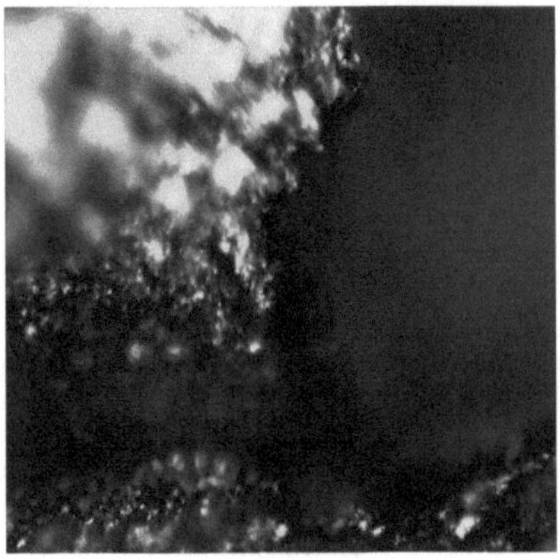

(c)

Fig. 5.38: Chips for aluminum alloy, CNT and Silicon carbide composite at, a) 80x magnification, b) 160x magnification, c) 320x magnification.

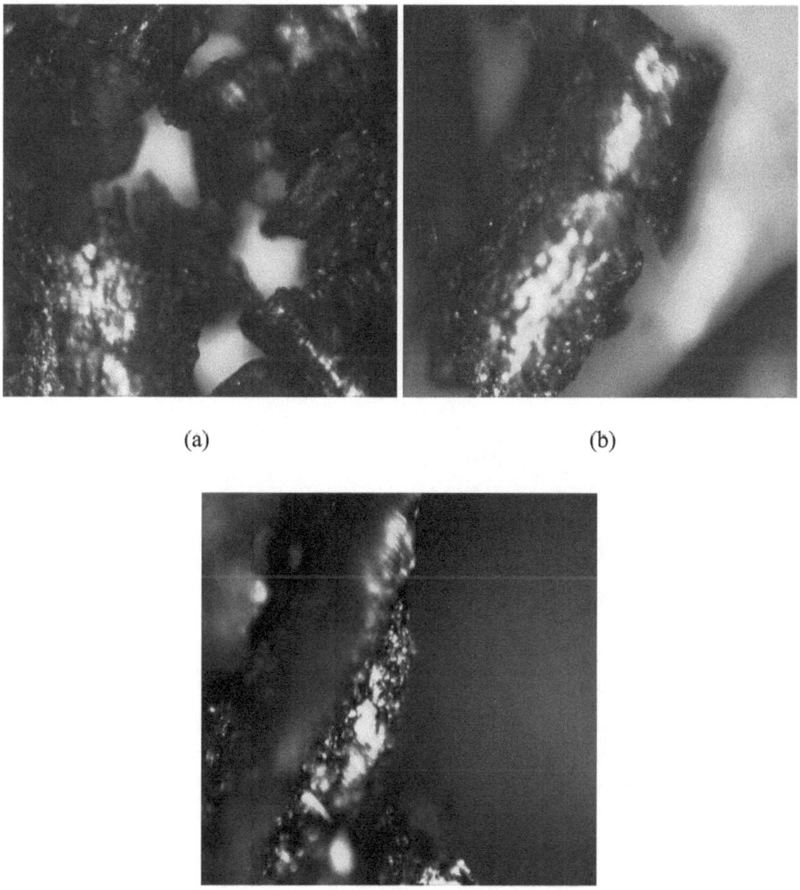

(a) (b)

(c)

Fig. 5.39: Chips for aluminum alloy, CNT, Silicon carbide and alumina composite at, a) 80x magnification, b) 160x magnification, c) 320x magnification.

Alloy and composites metal chips created during the dry machining process at 2260RPM and 0.2 mm/rev feed rate were collected and observed. The microscopic images are shown in Fig. 5.36 to Fig 5.39. From the figures it is observed that, for aluminum alloy the chips are darker compared to composite chips, which indicates the low heat transfer rate (Gururaj et al. 2021). From the images, the size of the chips and surface finish of the chips, as the percent reinforcement increased in the matrix material, the size of the chips getting smaller and for alloy the edges are more ragged compared to the composites, that indicates the delamination for aluminum alloy is higher than the composite materials.

CHAPTER 6
CONCLUSIONS AND RECOMMENDATIONS

6.1 Introduction

In the present work, a hybrid particle-reinforced aluminum metal matrix composite was fabricated using varying percentages of carbon nanotube, silicon carbide, and alumina through the stir casting process. The study aimed to evaluate the effects of these reinforcements on the composite's microstructure, physical, mechanical, and machining quality. The findings indicate that the Preference Selection Index (PSI) method effectively resolves conflicts in selecting suitable materials, confirming aluminum as an appropriate matrix material. Additionally, TOPSIS and MOORA methods are reliable for choosing reinforcing materials when data variation is low, with ranking stability linked to the initial criterion weight. Stir casting proved to be the preferred fabrication method, achieving finer reinforcement dispersion and fewer particle clusters compared to other techniques.

6.2 Conclusions

The conclusions are as follows:

- The microstructural analysis revealed a uniform dispersion of carbon nanotubes, silicon carbide, and alumina particles in the matrix phase. The tri-particle reinforced composite displayed better reinforcement distribution, although some clustering and porosity were noted.
- The addition of reinforcements reduced the density up to 0.83% in tri-particle reinforced composites compared to base alloy. Porosity increased significantly, with a 238.36% rise for the tri-particle reinforced composite relative to the base alloy.
- Reinforcements enhanced the mechanical properties. The bi-particle composite had the highest UTS (142.857%) and yield strength (143.75%) improvement compared to the base alloy. The tri-particle composite also showed increased UTS (128.57%) and yield strength (93.75%), though slightly lower than the bi-particle variant.
- Hardness increased with reinforcement addition, showing significant improvement compared to the base alloy. The inclusion of carbon nanotubes enhanced

impact strength by 100% over the base alloy. However, the addition of silicon carbide and alumina in tri-particle reinforced composite reduce the impact strength

- During dry machining, reinforced composites produced shorter, sharper-edged chips compared to that of the base alloy. This indicated reduced delamination and better machinability for the composites. Optimal parameters for tri-particle reinforced composites were 2260 rpm and 0.2 mm/rev feed rate, providing superior machining outcomes.

6.3 Recommendations

The recommendation of this present study is as follow:

- Further investigation can be carried out by varying reinforcement weight and machining parameters, such as depth of cut and wet machining, to study their effects on composite characteristics.
- Additional factors like production cost, machining time, and constraints such as machine power and material removal rate can be taken into consideration in optimization process to get more precious results.
- The stir casting process can be improved with speed and temperature control in a controlled environment.
- Investigations using other fabrication methods such as powder metallurgy, 3D printing etc. for hybrid composite development and comparing their physical, mechanical, and tribological properties with those of stir-casted composites would provide valuable insights for future advancements.

REFERENCES

Akinwamide, S. O., Akinribide, O. J., and Olubambi, P. A. (2021). Influence of ferrotitanium and silicon carbide addition on structural modification, nanohardness and corrosion behaviour of stir-cast aluminium matrix composites, Silicon, Vol 12, pp. 2221-2232.

Akshay, B. R., Keshavamurthy, R., Kuppahalli, P., and Sudhan, J. M. (2018). Mechanical properties of friction stir processed Al6061-BN surface composite, Materials Today: Proceedings, Vol. 5, Issue 11, pp. 24568–24577.

Partheeban, C. M. A., Rajendran, M., Vettivel, S. C., Suresh, S., and Moorthi, N. S. V. (2015). Mechanical behavior and failure analysis using online acoustic emission on nano-graphite reinforced Al6061-10TiB_2 hybrid composite using powder metallurgy, Materials Science and Engineering A, Vol. 632, pp. 1–13.

Rajesh, M. A., Kumar, S. S. B., Ashok, R., and Sekar, K. S. V. (2021). Experimental investigations on the drilling of titanium metal matrix composite, IOP Conference Series: Materials Science and Engineering, Vol. 1128, Issue 1.

Kumar, R. A., Karthick, K., Jayasuriya, R., Kumar, J. A., and Karthik, V. B. (2020). Tribological Behaviour of Aluminum Metal Matrix Composites - A Review, IOP Conference Series: Materials Science and Engineering, Vol. 923, Issue 1.

Bhat, A., and Kakandikar, G. (2019). Manufacture of silicon carbide reinforced aluminium 6061 metal matrix composites for enhanced sliding wear properties, Manufacturing Review, Vol. 6, pp. 4–9.

Bhoi, N. K., Singh, H., and Pratap, S. (2020). Developments in the aluminum metal matrix composites reinforced by micro/nano particles – A review', Journal of Composite Materials, Vol. 54, Issue 6, pp. 813–833.

Bhushan, R. K. (2020). Optimization of machining parameters for minimising cutting forces during machining of Al alloy SiC particle composites, Australian Journal of Mechanical Engineering, Vol. 20, Issue 2, pp. 372–386.

Bodunrin, M. O., Alaneme, K. K., and Chown, L. H. (2015). Aluminum matrix hybrid composites: A review of reinforcement philosophies; Mechanical, corrosion and tribological characteristics, Journal of Materials Research and Technology, Vol. 4, Issue 4, pp. 434–445.

Bragaglia, M., Montanari, R., and Montesperelli, G. (2019). Effect of Al_2O_3 reinforcement and precipitates on corrosion behaviour of 2618 and 6061 aluminium MMCs, Corrosion Engineering Science and Technology, Vol. 54, Issue 7, pp. 601–613.

Callister, W. D. (1991). Materials science and engineering: An introduction, Materials & Design, Vol. 12, Issue 1.

Callister, W. D. D. (2019). Materials science and engineering: An introduction, 10th australia and New Zealand edition, Materials & Design.

Canakci, A., Varol, T., and Ertok, S. (2012). The effect of mechanical alloying on Al_2O_3 distribution and properties of Al_2O_3 particle reinforced Al-MMCs, Science and Engineering of Composite Materials, Vol. 19, Issue 3, pp. 227–235.

Chakrapani, P., and Suryakumari, T. S. A. (2021). Mechanical properties of aluminium metal matrix composites-A review, Materials Today: Proceedings, Vol. 45, pp. 5960–5964.

Chen, Y. C., and Ou, S. F. (2020). Effects of reinforcement ratios and sintering temperatures on the mechanical properties of titanium nitride/nickel composites, Materials, Vol. 13, Issue 20, pp. 1–12.

Devanathan, R., Ravikumar, J., Boopathi, S., Selvam, D. C., and Anicia, S. A. (2019). Influence in mechanical properties of stir cast aluminum (AA6061) hybrid metal matrix composite (HMMC) with Silicon Carbide, Fly Ash and Coconut coir Ash reinforcement, Materials Today: Proceedings, Vol. 22, pp. 3136–3144.

Dhanashekar, P., Loganathan, S., Ayyanar, S. R., Mohan, T., and Sathish, M. (2020). Obtaining hydrophobic surfaces in atmospheric pressure plasma, Materials Today: Proceedings, Vol. 20, pp. 335–341.

El-Kady, O., and Fathy, A. (2014). Effect of SiC particle size on the physical and mechanical properties of extruded Al matrix nanocomposites, Materials and Design, Vol. 54, pp. 348–353.

Elumalai, N., Karthik, M. K., Giridharan, V., and Kumar, C. S. (2020). Mechanical properties of titanium carbide reinforced aluminum metal matrix composites, AIP Conference Proceedings, Vol. 2271.

Emiru, A. A., Sinha, D. K., Kumar, A., and Yadav, A. (2022). Fabrication and Characterization of Hybrid Aluminum (Al 6061) Matrix Composite Reinforced with SiC , B_4C and MoS_2 via stir casting, International Journal of Metalcasting, Vol. 17.

Escaich, C., Shi, Z., Baron, L., and Balazinski, M. (2020). Machining of titanium metal matrix composites: Progress overview, Materials, Vol. 13, Issue 21, pp. 1–18.

Kuwarmausam, Singh, P. P., and Mishra, A. (2021). Experimental Evalutionof tensile strength of cryogenically solidified aluminum based metal matrix composites (MMC), Turkish Journal of Computer and Mathematics Education (TURCOMAT), Vol. 12, Issue 2, pp. 2053–2059.

Edoziuno, F. O., Nwaeju, C. C., Adediran, A. A., Odoni, B. U., and Prakash, A. V. R. (2021). Mechanical and microstructural characteristics of aluminum 6063 alloy/palm kernel shell composites for lightweight applications, Scientific African, Vol 12,

Gangadharappa, M., Reddappa, H. N., Ravikumar, M., and Suresh, R. (2018). Mechanical and wear characterization of Al6061 red mud composites, Materials Today: Proceedings, Vol. 5, Issue 10, pp. 22384–22389.

Garg, P., Jamwal, A., Kumar, D., Sadasivuni, K. K., Hussain, C. M., and Gupta, P. (2019). Advance research progresses in aluminium matrixcomposites: Manufacturing & applications, Journal of Materials Research and Technology, Vol. 8, Issue 5, pp. 4924-4939.

Gowri Shankar, M. C., Shettar, M., Sharma, S. S., Kini, A., and Jayashree (2018). Enhancement in hardness and influence of artificial aging on stir cast Al6061-B4C and Al6061-SiC composites, Materials Today: Proceedings, Vol. 5, Issue 1, pp. 2435–2443.

Gupta, M. (2018). Metal matrix composites, Metals, Vol. 8, Issue 6.

Halil, K., Ovali, I., Dundar, S., and Citak, R. (2019). Wear and mechanical properties of Al6061/SiC/B_4C hybrid composites produced with powder metallurgy, Journal of Materials Research and Technology, Vol. 8, Issue 6, pp. 5348–5361.

Hashim, H., Salleh, M. S., Omar, M. Z., Sulong, A. B., and Rahman, A. A. (2021). Influence of short heat treatment on the microstructures and mechanical properties of Thixoformed aluminum alloy composite, Jurnal Tribologi, Vol. 28, pp. 96–104.

Idrisi, A. H., Mourad, A. H. I., Thekkuden, D. T., and Christy, J. V. (2018). Wear behavior of AA 5083/SiC nano-particle metal matrix composite: Statistical analysis, IOP Conference Series: Materials Science and Engineering, Vol. 324, Issue 1.

Idrisi, A. H., and Mourad, A. H. I. (2019). Conventional stir casting versus ultrasonic assisted stir casting process: Mechanical and physical characteristics of AMCs, Journal of Alloys and Compounds, Vol. 805, pp. 502–508.

Jamaati, R., Amirkhanlou, S., Toroghinejad, M. R., and Niroumand, B. (2011). Effect of particle size on microstructure and mechanical properties of composites produced by ARB process, Materials Science and Engineering A, Vol. 528, Issue 4–5, pp. 2143–2148.

Kamboj, A., Kumar, S., and Singh, H. (2013). Design and development of hybrid stir casting process, International Journal of Applied Industrial Engineering, Vol. 1, Issue 2, pp. 1–6.

Kandpal, B. C., Kumar, J., and Singh, H. (2017). Fabrication and characterisation of Al2O3/aluminium alloy 6061 composites fabricated by Stir casting, Materials Today: Proceedings, Vol. 4, Issue 2, pp. 2783–2792.

Kareem, A., Qudeiri, J. A., Abdudeen, A., Ahammed, T., and Ziout, A. (2021). A review on AA 6061 metal matrix composites produced by stir casting, Materials, Vol. 14, Issue 1, pp. 1–22.

Kishore, D. S. C., Rao, K. P., Basha, S. M. J., and Rao, B. J. P. (2018). Investigation of surface roughness in turning of in-situ Al6061-TiC metal matrix composite by taguchi and prediction of response by ANN, Materials Today: Proceedings, Vol. 5, Issue 9, pp. 18070–18079.

Jaykumar, M. S., and Kohir, V. V. (2020). Microstructure, hardness and chip formation of an aluminum alloy metal matrix composite fabricated by squeeze casting method, Vol. 7, Issue 7. pp. 37-41.

Kumar, J., Singh, D., Kalsi, N. S., Sharma, S., Pruncu, C. I., Pimenov, D. Y., Rao, K. V., and Kaplonek, W. (2020). Comparative study on the mechanical, tribological, morphological and structural properties of vortex casting processed, Al-SiC-Cr hybrid metal matrix composites for high strength wear-resistant applications: Fabrication and characterizations, Journal of Materials Research and Technology, Vol. 9, Issue 6, pp. 13607–13615.

Kumar, V. A., Anil, M. P., Rajesh, G. L., Hiremath, V., and Auradi, V. (2018). Tensile and compression behavior of boron carbide reinforced 6061 Al MMC's processed through conventional melt stirring, Materials Today: Proceedings, Vol. 5, Issue 8, pp. 16141–16145.

Kumar, V., Gupta, R. D., and Batra, N. K. (2014). Comparison of mechanical properties and effect of sliding velocity on wear properties of Al 6061, Mg 4%, Fly Ash and Al 6061, Mg 4%, Graphite 4%, Fly Ash hybrid metal matrix composite, Procedia Materials Science, Vol. 6, pp. 1365–1375.

Madhukumar and Umashankar (2018). Characterization of Glass Particulate Reinforced Aluminium Alloy6061 Metal Matrix Composites, Materials Today: Proceedings, Vol. 5, Issue 2, pp. 7604–7608.

Manigandan, K., Srivatsan, T. S., Ren, Z., and Zhao, J. (2015). Influence of reinforcement content on tensile response and fracture behavior of an aluminum alloy metal matrix composite, Advanced Composites for Aerospace, Marine, and Land Applications II, (April), pp. 101–119.

Gass, S. I., and Fu, M. C. (2013). Encyclopedia of operations research and management science, Encyclopedia of Operations Research and Management Science. doi: 10.1007/978-1-4419-1153-7.

Maurya, N. K., Maurya, M., Srivastava, A. K., Dwivedi, S. P., Kumar, A., and Chauhan, S. (2019). Investigation of mechanical properties of Al 6061/SiC composite prepared through stir casting technique, Materials Today: Proceedings, Vol. 25, Part 4, pp. 755–758.

Mavhungu, S. T., Akinlabi, E. T., Onitiri, M. A., and Varachia, F. M. (2017). Aluminum matrix composites for industrial use: advances and trends, Procedia Manufacturing, Vol. 7, pp. 178–182.

Nagaral, M., Auradi, V., Parashivamurthy, K. I., Kari, S. A., and Shivananda, B. K. (2018). Synthesis and characterization of Al6061-SiC-graphite composites fabricated by liquid metallurgy, Materials Today: Proceedings, Vol. 5, Issue 1, pp. 2836–2843.

Nanjan, S., and Janakiram, G. M. (2019). Characteristics of A6061(Glass Fibre + AL_2O_3 + SiC + B_4C) reinforced hybrid composite prepared through stir casting, Advances in Materials Science and Engineering, Vol. 5, pp. 1-12.

Nusrat Tarin Chowdhury (2011). Optimization of cutting parameters in near dry machining of hardened steel, Thesis, Dept. of IPE, Bangladesh university of engineering & technology (BUET).

Oladijo, O. P., Awe, S. A., Akinlabi, E. T., Phiri, R. R., Collieus, L. L., and Phuti, R. E. (2021). High-temperature properties of metal matrix composites, Reference Module in Materials Science and Materials Engineering, (January), Vol. 1, pp. 1–15.

Pitchayyapillai, G., Seenikannan, P., Balasundar, P., and Narayansamy, P. (2017). Effect of nano-silver on microstructure, mechanical and tribological properties of cast 6061 aluminum alloy, Transactions of Nonferrous Metals Society of China (English Edition), Vol. 27, Issue 10, pp. 2137–2145.

Prabhu, S. R., Shettigar, A. K., Herbert, M. A., and Rao, S. S. (2019). Microstructure and mechanical properties of rutile-reinforced AA6061 matrix composites produced via stir casting process, Transactions of Nonferrous Metals Society of China (English Edition), Vol. 29, Issue 11, pp. 2229–2236.

Prakash, M., Badhotiya, G. K., and Chauhan, A. S. (2019). A review on mechanical and wear characteristics of particulate reinforced Al-alloy based MMC, AIP Conference Proceedings, Vol. 2148.

Kumar, P. H. G., and Xavior, A. M. (2017). Processing and characterization of Al 6061 - graphene nanocomposites, Materials Today: Proceedings, Vol. 4, Issue 2, pp. 3308–3314.

Pravin, R., and Raj, S. K. (2019). Stir casting & Processing of Aluminum Matrix Composites', AIP Conference Proceedings, Vol. 2142.

Premnath, A. A., Alwarsamy, T., Rajmohan, T., and Prabhu, R. (2014). The influence of alumina on mechanical and tribological characteristics of graphite particle reinforced hybrid Al-MMC, Journal of Mechanical Science and Technology, Vol. 28, Issue 11, pp. 4737–4744.

Ramanathan, A., Krishnan, P. K., and Muraliraja, R. (2019). A review on the production of metal matrix composites through stir casting – Furnace design, properties, challenges, and research opportunities, Journal of Manufacturing Processes, Vol. 42, pp. 213–245.

Ravi, K. R., Sreekumar, V. M., Pillai, R. M., Mahato, C., Amaranathan, K. R., Kumar, R. A., and Pai, B. C. (2007). Optimization of mixing parameters through a water model for metal matrix composites synthesis, Materials and Design, Vol. 28, Issue 3, pp. 871–881.

Ravi, K., Kiran, K., and Sreebalaji, V. S. (2017). Micro structural characteristics and mechanical behaviour of aluminium matrix composites reinforced with titanium carbide, Journal of Alloys and Compounds, Vol. 723, pp. 795–801.

Ravikumar, M., Reddappa, H. N., Suresh, R., Reddy, M., Babu, E., Reddy, N. C., Ravikumar, C., and Murthy, H. (2021). Evaluation of corrosion properties of Al_2O_3 and SiC reinforced aluminium metal matrix composites using Taguchi's techniques, Journal of scientific research, Vol. 65, Issue 1, pp. 253–259.

Ravindran, S., Mani, N., Balaji, S., Abhijith, M., and Surendaran, K. (2019). Mechanical behaviour of aluminium hybrid metal matrix composites - A review, Materials Today: Proceedings, Vol. 16, Part 2, pp. 1020–1033.

Rebba, B., and Ramanaiah, N. (2014). Evaluation of mechanical properties of aluminium alloy (al-2024) reinforced with molybdenum disulphide (MOS_2) metal matrix composites, Procedia Materials Science, Vol. 6, pp. 1161–1169.

Romanova, V. A., Balokhonov, R. R., and Schmauder, S. (2009) The influence of the reinforcing particle shape and interface strength on the fracture behavior of a metal matrix composite, Acta Materialia, Vol. 57, Issue 1, pp. 97–107.

Rouhi, M., Moazami-Goudarzi, M., and Ardestani, M. (2019). Comparison of effect of SiC and MoS_2 on wear behavior of Al matrix composites, Transactions of Nonferrous Metals Society of China (English Edition), Vol. 29, Issue 6, pp. 1169–1183.

Sagar, K. G., and Suresh, P. M. (2020) Study on mechanical properties and microstructure of particle reinforced al2024-beryl composites by cold, Vol. 11, pp. 44–48.

Saheb, N., Khalil, A., Hakeem, A. S., Laoui, T., and Al-Aqeeli, N. (2012). Carbon nanotube reinforced Al6061 and Al2124 nanocomposites, ECCM 2012 - Composites at Venice, Proceedings of the 15th European Conference on Composite Material-2012.

Sahu, K., Rana, R. S., Purohit, R., Koli, D. K., Rajpurohit, S. S., and Singh, M. (2015). Wear behavior and micro-structural study of al/al_2o_3nano-composites before and after heat treatment, Materials Today: Proceedings, Vol. 2, Issue 4–5, pp. 1892–1900.

Saranu, R., Chanamala, R., and Putti, S. (2020). Review of magnesium metal matrix composites, IOP Conference Series: Materials Science and Engineering, Vol. 961, Issue 1.

Satheesh, M., and Pugazhvadivu, M. (2019). Investigation on physical and mechanical properties of Al6061-Silicon Carbide (SiC)/Coconut shell ash (CSA) hybrid composites, Physica B: Condensed Matter, Vol. 572, pp. 70–75.

Selvam, D. R. J., and Dinaharan, D. R. I. (2013). Synthesis and characterization of Al6061-Fly Ashp-SiCp composites by stir casting and compocasting methods, Energy Procedia, Vol. 34, pp. 637–646.

Senapathi, K., Raju, S. S., and Rao, G. S. (2017). Tribological Performance of Al-MMC Reinforced withTreated Fly Ash using Response Surface Methodology, Indian Journal of Science and Technology, Vol. 10, Issue 15, pp. 1–9.

Senapati, A. K., Panda, S. S., Dutta, B. K., and Mishra, S. (2020). Effect of stirring speed during casting on mechanical properties of Al–Si Based MMCs, Smart Innovation, Systems and Technologies. Springer singapore. pp. 703-710.

Sharma, A., Sharma, V. M., and Paul, J. (2019). A comparative study on microstructural evolution and surface properties of graphene/CNT reinforced Al6061−SiC hybrid surface composite fabricated via friction stir processing', Transactions of nonferrous metals society of china (english edition), Vol. 29, Issue 10, pp. 2005–2026.

Sharma, V. K., Kumar, V., and Joshi, R. S. (2019) 'Investigation of rare earth particulate on tribological and mechanical properties of Al-6061 alloy composites for aerospace application', Journal of Materials Research and Technology, Vol. 8, Issue 4, pp. 3504–3516.

Shimizu, H., Dolowy Jr., J. F., Taylor, R. J., and Webb, B. A. (1967). Metal-matrix composites behavior and aerospace applications, SAE Technical, Vol. 76, Section 4, pp. 2661-26668.

Siddesha, S. (2016). Effects of fabrication of aluminium 2024 / TiO_2 metal matrix composite, International Journal of Innovative Research & Development, Vol. 5, Issue 11, pp. 174–177.

Singh, H., Singh, G., Singh, K., and Vardhan, S. (2021). Evaluation of mechanical performance on a developed AA 6061matrix-Mg/0.9-Si/0.68 reinforced with B4C based composites, Functional Composites and Structures, Vol. 3, Issue 1, pp. 0–7.

Singh, J., and Chauhan, A. (2016). Characterization of hybrid aluminum matrix composites for advanced applications - A review, Journal of Materials Research and Technology, Vol. 5, Issue 2, pp. 159–169.

Sivananthan, S., Ravi, K. and Samuel, S. J. (2020). Effect of SiC particles reinforcement on mechanical properties of aluminium 6061 alloy processed using stir casting route, Materials Today: Proceedings, Vol. 21, Part 1, pp. 968–970.

Slipenyuk, A., Kuprin, V., Milman, Yu., Goncharuk, V., and Eckert, J. (2006). Properties of P/M processed particle reinforced metal matrix composites specified by reinforcement concentration and matrix-to-reinforcement particle size ratio, Acta Materialia, Vol. 54, Issue 1, pp. 157–166.

Akula, S., Nayak, S. N., Bolar, G., and Managuli, V. (2021). Comparison of conventional drilling and heilical milling for hole making in Ti_6Al_4V titanium alloy under sustainable dry condition', Manufacturing Rev. 8, vol. 12.

Song, S. G., Shi, N., Gray III, G. T., and Roberts, J. A. (1996). Reinforcement shape effects on the fracture behavior and ductility of particulate-reinforced 6061-Al matrix composites. Metallurgical and Materials Transactions A, Vol. 27, Issue 11, pp. 3739–3746.

Sucharitha, M. and Ravisankar, B. (2018). Preparation and characterization of aluminium metal matrix composites, Advanced Science, Engineering and Medicine, Vol. 10, Issue 3, pp. 304–307.

Iijima, S. (1991). Synthesis of carbon nanotubes, Nature, Vol. 354, pp. 56–58.

Suresh, S., Moorthi, N. S. V., Vettivel, S. C., Selvakumar, N., and Jinu, G. R. (2014). Effect of graphite addition on mechanical behavior of Al6061/TiB2 hybrid composite using acoustic emission, Materials Science and Engineering A, Vol. 612, pp. 16–27.

Tamizharasan, T., Senthilumar, N., Selvakumar, V., and Dinesh, S. (2019). Taguchi's methodology of optimizing turning parameters over chip thickness ratio in machining P/M AMMC, SN Applied Sciences, Vol. 1, No 160.

Toor, Z. S. (2017). Applications of Aluminum-Matrix Composites in Satellite: A Review', Journal of Space Technology, Vol. 7, Issue 1, p. 6.

Venkatesan, S., and Xavior, M., A. (2019). Wear property evaluation of aluminum alloy (aa7050) metal matrix composite reinforced with graphene fabricated by stir and squeeze cast processes, Materials Today: Proceedings, Vol. 22, Part 4, pp. 3330–3339.

Verma, V., Kumar, P., Mittal, K., Chauhan, S., and Tewari, P. (2016). Microstructure and mechanical behaviour characterisation of Al-Al2O3 MMC processed by DIMOX and Al-Al_2O_3/MnO_2 MMC processed via stir casting route, International Journal of Materials Engineering Innovation, Vol. 7, Issue 3–4, pp. 219–235.

Votarikari, N., Nath, N. K., and Babu, P. R. (2022). Influence of nano-silica in al6061 composites on mechanical and material properties prepared through powder metallurgy, Advance in Materials and Processing Technologies, Vol. 8, Issue 1, pp. 478-497.

Williams, J. J., Flom, Z., Amell, A. A., Chawla, N., Xiao, X., and Carlo, F. De. (2010). Damage evolution in SiC particle reinforced Al alloy matrix composites by X-ray synchrotron tomography, Acta Materialia, Vol. 58, Issue 18, pp. 6194–6205.

Yang, Z., Fan, Z., Liu, Y., Nie, J., Yang, Z., and Kang, Y. (2021). Effect of the particle size and matrix strength on strengthening and damage process of the particle reinforced metal matrix composites, Materials, Vol. 14, Issue 3, pp. 1–12.

Yashpal, Jawalkar, C. S., Kant, S., Panwar, N., Sharma, M. D., and Pali, H. S. (2020). Effect of particle size variation of bagasse ash on mechanical properties of aluminium hybrid metal matrix composites, Materials Today: Proceedings, Vol. 21, Part 4, pp. 2024–2029.

Yeshiye, T. and Gizaw, M. (2021). A review on Effects of reinforcements on properties and wear behaviour of aluminium metal matrix material, Vol. 6, Issue 2, pp. 1–17.

Youssef, Y. M., and El-Sayed, M. A. (2016). Effect of reinforcement particle size and weight fraction on the mechanical properties of SiC particle reinforced al metal matrix composites, International Review of Mechanical Engineering, Vol. 10, Issue 4, pp. 261–265.

Zhang, Z., Chen, D. L. (2008). Contribution of Orowan strengthening effect in particulate-reinforced metal matrix nanocomposites, Mater. Sci. Eng. A, Vol. 483-484, pp. 148-152.

Zhang, Z., Chen, D. L. (2006). Consideration of Orowan strengthening effect in particulate reinforced metal matrix nanocomposites: a model for predicting their yield strength, Scripta Materialia, Vol. 54, Issue 7, pp. 1321-1326.

APPENDICES

Appendix – A: Matrix Selection Using PSI (Preference Selection Index) Method

To evaluate the criteria weights using PSI method following steps were followed:

Step 1: Defining the goal of the analysis and breaking down the MCDM (multi-criteria decision making) into aim, criteria and alternative.

The aim of the analysis was to select a matrix material best suitable for the applications in industrial area. The criteria for better performances were higher shear strength, yield strength, hardness, young's modulus, thermal conductivity and lower density, melting points, specific heat. The alternatives for this study were the twenty-two metals and alloy, namely iron, aluminum, nickel, beryllium, copper, magnesium, manganese, molybdenum, tin, titanium, tungsten, molybdenum alloy, cobalt alloy, aluminum alloy, beryllium alloy, magnesium alloy, nickel alloy, tin alloy, zinc alloy, tungsten alloy, alloy cast iron, titanium alloy.

Step 2: Formulate the decision matrix. The solving of each multi-criteria decision making (MCDM) problem begins with constructing decision matrix. Table A.1 shows the data of alternatives.

Table A.1: Decision matrix for alternative matrix materials

SN	Materials	Shear strength, MPa	Yield tensile strength, MPa	Hardness, Brinell	Young's modulus, GPa	Density, g/cc	Melting point, deg C	Thermal conductivity, W/m-K	Specific heat, J/g deg C
1	Fe, iron	358	671	251	196	7.79	1440	16.7	0.478
2	Aluminum, AL	207	276	95	68.9	2.7	652	167	0.896
3	Nickel	723	590	282	207	8.4	1320	16.7	0.431
4	Beryllium	345	240	85	303	1.844	1283	216	1.925
5	Copper	210	33.3	50	110	7.764	1083	385	0.385
6	Magnesium, Mg	165	105	35	44	1.74	649.3	159	1.025

Table A.1: Continued

SN	Materials	Shear strength, MPa	Yield tensile strength, MPa	Hardness, Brinell	Young's modulus, GPa	Density, g/cc	Melting point, deg C	Thermal conductivity, W/m-K	Specific heat, J/g deg C
7	Manganese, Mn	496	241	460	159	7.44	1244	7.82	0.448
8	Molybdenum, Mo	500	324	225	330	10.22	2617	138	0.255
9	Tin, Sn	220	11	2.3	41.6	5.765	231.968	63.2	0.256
10	Titanium, Ti	220	140	70	116	4.5	1670	17	0.528
11	Tungsten, W	400	750	294	400	19.3	3370	163.3	0.134
12	Molybdenum alloy	500	567	181	303	10	2340	173	0.255
13	Cobalt alloy	415	760	323	191	8.4	1350	14.7	0.423
14	Aluminum alloy	165	180	84.5	72	2.73	579	135	0.962
15	Beryllium alloy	298	672	256	155	7.57	1010	150	1
16	Magnesium alloy	145	164	63.7	45.2	1.81	540	91.1	1.02
17	Nickel alloy	723	590	282	207	8.4	1320	16.7	0.431
18	Tin alloy	39.7	33	16.4	37.4	7.8	203	46.2	0.211
19	Zinc alloy	225	277	72.1	83.2	6.61	424	109	0.416
20	Tungsten alloy	394	678	925	340	15.4	2110	155	0.172
21	Alloy cast iron	712	520	262	156	7.17	1150	26.6	0.506
22	Titanium alloy, general	608	778	305	114	4.5	1650	7.64	0.519

Step 3: The data normalization.

The process of transforming attributes value into a range of 0-1 is called normalization, and it is required in multi attribute decision matrix making method to transform performance rating with different data measurement unit in a decision matrix into a compatible unit. If the expectancy is the larger the better (i.e. profit), then the original attribute performance can be normalized as follows:

$$R_{ij} = \frac{x_g}{x_j^{max}} \quad \text{(A.1)}$$

If the expectancy is the smaller the better (i.e. cost), then the original attributes performance can be normalized as follows:

$$R_{ij} = \frac{x_j^{min}}{x_g} \quad \text{(A.2)}$$

The following Table A.2 shows the normalization matrix.

Table A.2: Normalized matrix

SN	Materials	Shear strength, MPA	Yield tensile strength, MPa	Hardness, Brinell	Young's modulus,Gpa	Density, g/cc	Melting point, deg C	Thermal conductivity, W/m-K	Specific heat, J/g degc
1	Fe, iron	0.495	0.862	0.271	0.49	0.223	0.141	2.186	0.280
2	Aluminum, AL	0.286	0.355	0.103	0.172	0.644	0.311	21.858	0.150
3	Nickel	1	0.759	0.305	0.518	0.207	0.154	2.186	0.311
4	Beryllium	0.477	0.308	0.092	0.758	0.944	0.158	28.272	0.070
5	Copper	0.290	0.043	0.054	0.275	0.224	0.187	50.393	0.348
6	Magnesium, Mg	0.228	0.135	0.038	0.11	1	0.313	20.811	0.131

Table A.2: Continued

SN	Materials	Shear strength, MPA	Yield tensile strength, MPa	Hardness, Brinell	Young's modulus,Gpa	Density, g/cc	Melting point, deg C	Thermal conductivity, W/m-K	Specific heat, J/g degc
7	Manganese, Mn	0.687	0.311	0.498	0.398	0.234	0.163	1.023	0.299
8	Molybdenum, Mo	0.691	0.416	0.243	0.825	0.170	0.078	18.063	0.525
9	Tin, Sn	0.304	0.014	0.002	0.104	0.303	0.875	8.272	0.523
10	Titanium, Ti	0.304	0.181	0.075	0.29	0.386	0.122	2.225	0.254
11	Tungsten, W	0.553	0.964	0.318	1	0.090	0.060	21.374	1
12	Molybdenum alloy	0.692	0.729	0.196	0.758	0.174	0.087	22.644	0.525
13	Cobalt alloy	0.574	0.977	0.349	0.477	0.207	0.150	1.924	0.316
14	Aluminum alloy	0.228	0.2316	0.091	0.18	0.637	0.350	17.670	0.139
15	Beryllium alloy	0.412	0.863	0.276	0.388	0.231	0.201	19.633	0.134
16	Magnesium alloy	0.201	0.211	0.068	0.113	0.961	0.376	11.924	0.132
17	Nickel alloy	1	0.758	0.305	0.518	0.207	0.154	2.186	0.311
18	Tin alloy	0.055	0.042	0.018	0.093	0.223	1	6.0471	0.635
19	Zinc alloy	0.311	0.356	0.078	0.208	0.263	0.478	14.267	0.322
20	Tungsten alloy	0.545	0.871	1	0.85	0.113	0.096	20.288	0.779
21	Alloy cast iron	0.985	0.668	0.283	0.39	0.242	0.177	3.482	0.265

Table A.2: Continued

SN	Materials	Shear strength, MPA	Yield tensile strength, MPa	Hardness, Brinell	Young's modulus,Gpa	Density, g/cc	Melting point, deg C	Thermal conductivity, W/m-K	Specific heat, J/g degc
22	Titanium alloy, general	0.841	1	0.331	0.285	0.387	0.123	1	0.258

Step 4: Compute preference variation value (PV_j)

In this step, preference variation value (PV_j) for each attributes are determined with concept of sample variance analogy using following equations:

$$PV_j = \sum_{i=1}^{N} \left(R_{ij} - \bar{R}_j\right)^2 \qquad (A.3)$$

Where, R_j is the mean of normalized value of attributes j and $\bar{R}_j = \frac{1}{N}\sum_{i=1}^{N} R_{ij}$

The following Table A.3 shows the preference variation value matrix.

Table A.3: preference variation value matrix

SN	Materials	Shear strength, MPA	Yield tensile strength, MPa	Hardness, Brinell	Young's modulus ,Gpa	Density, g/cc	Melting point, deg C	Thermal conductivity, W/m-K	Specific heat, J/g degc
1	Fe, iron	0.001	0.13	0.002	0.005	0.021	0.015	128.76	0.005
2	Aluminum, AL	0.049	0.022	0.015	0.060	0.077	0.002	69.313	0.041
3	Nickel	0.243	0.065	0.006	0.009	0.026	0.012	128.76	0.002
4	Beryllium	0.001	0.038	0.018	0.115	0.333	0.011	217.24	0.078
5	Copper	0.047	0.211	0.029	0.021	0.020	0.005	1358.62	5.371
6	Magnesium, Mg	0.078	0.135	0.036	0.095	0.401	0.003	52.974	0.048

Table A.3: Continued

SN	Materials	Shear strength, MPA	Yield tensile strength, MPa	Hardness, Brinell	Young's modulus, Gpa	Density, g/cc	Melting point, deg C	Thermal conductivity, W/m-K	Specific heat, J/g degc
7	Manganese, Mn	0.032	0.037	0.073	0.001	0.018	0.011	156.49	0.0027
8	Molybdenum, Mo	0.034	0.007	0.001	0.166	0.039	0.034	20.518	0.0307
9	Tin, Sn	0.041	0.238	0.050	0.099	0.004	0.376	27.678	0.031
10	Titanium, Ti	0.041	0.104	0.023	0.016	0.001	0.021	127.87	0.0093
11	Tungsten, W	0.002	0.213	0.008	0.339	0.077	0.041	61.484	0.422
12	Molybdenum alloy	0.034	0.051	0.001	0.115	0.037	0.031	83.006	0.031
13	Cobalt alloy	0.004	0.225	0.015	0.004	0.026	0.012	134.771	0.001
14	Aluminum alloy	0.078	0.074	0.018	0.057	0.073	0.008	17.114	0.044
15	Beryllium alloy	0.009	0.131	0.002	0.001	0.019	0.004	37.214	0.047
16	Magnesium alloy	0.094	0.085	0.025	0.093	0.353	0.013	2.589	0.048
17	Nickel alloy	0.243	0.065	0.006	0.011	0.026	0.012	128.762	0.002
18	Tin alloy	0.205	0.212	0.044	0.105	0.021	0.545	56.041	0.081
19	Zinc alloy	0.038	0.021	0.022	0.044	0.011	0.047	0.538	0.001
20	Tungsten alloy	0.001	0.136	0.597	0.187	0.064	0.027	45.627	0.184
21	Alloy cast iron	0.228	0.028	0.003	0.001	0.015	0.007	101.033	0.007
22	Titanium alloy, general	0.111	0.248	0.011	0.018	0.001	0.019	157.081	0.0089

Step 5: Determine overall preference value ψ_j

In this step, the overall preference value is determined for each attributes. To get the overall preference value, it is required to find deviation ϕ_j in preference value PV_j and the deviation in preference value for each attributes is determined using the following equation:

$$\phi_j = 1 - PV_j \tag{A.4}$$

And overall preference value ψ_j is determined using following equation:

$$\psi_j = \frac{\phi_j}{\sum_{j=1}^{M} \phi_j} \tag{A.5}$$

The total overall preference value of all the attributes should be one, i.e. $\sum_{j=1}^{M} \psi_j = 1$.

Step 6: Obtain the preference selection index I_j

Now, compute the preference selection index I_j for each alternative using the following equation.

$$I_j = \sum_{j=1}^{M} (R_{ij} \times \psi_j) \tag{A.6}$$

Step 7: After calculation of the preference selection index, alternatives are ranked according to ascending or descending order to facilitate the managerial interpretation of results.

Table A.4: Final alternative selection

SN	Materials	Preference selection index	Ranking
1	Fe, iron	2.1839	19
2	Aluminum, AL	21.8331	4
3	Nickel	2.1839	17
4	Beryllium	28.2393	2

Table A.4: Continued

SN	Materials	Preference selection index	Ranking
5	Copper	50.3331	1
6	Magnesium, Mg	20.7872	6
7	Manganese, Mn	1.0228	21
8	Molybdenum, Mo	18.0419	9
9	Tin, Sn	8.2627	13
10	Titanium, Ti	2.2228	16
11	Tungsten, W	21.3498	5
12	Molybdenum alloy	22.6178	3
13	Cobalt alloy	1.9225	20
14	Aluminum alloy	17.6496	10
15	Beryllium alloy	19.6109	8
16	Magnesium alloy	11.9104	12
17	Nickel alloy	2.1840	17
18	Tin alloy	6.0402	14
19	Zinc alloy	14.2505	11
20	Tungsten alloy	20.2646	7
21	Alloy cast iron	3.4782	15
22	Titanium alloy, general	0.9996	22

From the table, we have seen that the aluminum ranked fourth in matrix. In this calculation process the availability of each material are not considered, as all of them are variable in market randomly. If we consider availability of material of top five ranked alternatives, aluminum is the material selected as matrix.

Appendix – B: Reinforcement Selection Using Technique for Order of Preference by Similarity to Ideal Solution (TOPSIS) Method

The TOPSIS analysis was used to determine the best reinforcements for developing hybrid particle reinforced metal matrix composites. The following stages were used in the process.

Step 1: Developing the decision matrix

The decision matrix for the TOPSIS analysis was formed applying the following equation, the alternatives or the variants of reinforcements are represented as A1, A2, …, Am. The criteria such as density, price, melting point, modulus of elasticity are represented as C1, C2, …, Cm. The values of each criterion are represented as Xij.

$$D = \begin{array}{c} \\ A_1 \\ A_2 \\ \vdots \\ A_m \end{array} \begin{bmatrix} C_1 & C_2 & \cdots & C_n \\ x_{11} & x_{12} & \cdots & x_{1n} \\ x_{21} & x_{22} & \cdots & x_{2n} \\ \vdots & \vdots & \vdots & \vdots \\ x_{m1} & x_{m2} & \cdots & x_{mn} \end{bmatrix} \tag{B.1}$$

The Table B.1 represents the decision matrix.

Table B.1: Decision matrix for alternative particulate reinforcements

Serial No	Name	Density g/cc	Price dollar (2021)	Melting point deg C	Modulus of elasticity GPa
1	Aluminum oxide Al_2O_3	3.96	6	2054	370
2	Barium boride BaB_6	4.35	20	2270	385
3	Chromium carbide Cr_3C_2	6.67	0.66	1800	380
4	Hafnium Carbide HfC	12.2	105	3000	352

Table B.1: Continued

Serial No	Name	Density g/cc	Price dollar (2021)	Melting point deg C	Modulus of elasticity GPa
5	Niobium Carbide NbC	7.6	80	3500	450
6	Silicon Carbide SiC	3.1	4.5	2797	420
7	Tantalum Carbide TaC	14.3	228	3900	550
8	Titanium Carbide, TiC	4.94	20	3065	450
9	Vanadium Carbide VC	5.71	45	2730	350
10	Zirconium Carbide ZrC	6.65	89	3400	390
11	Chromium nitride CrN	5.9	15	1050	83
12	CNT	1.35	2	3550	800
13	GNP	2.26	50	3550	210
14	Boron carbide B_4C	2.52	30	2350	289.65
15	Manganese dioxide MnO_2	5.026	50	535	276.34
16	Silicon nitride Si_3N_4	3.29	85	1000	310
17	Aluminum nitride AlN	3.2	9	1700	330
18	Titanium Nitride TiN	5.22	15	2950	600
19	Zirconium Oxide ZrO_2	5.7	20	2750	205
20	Molybdenum Disulfide MoS_2	5.06	30	1185	330

Step 2: Developing the normalized and fuzzy weighted normalized decision matrix.

The normalized decision matrix for the TOPSIS method and weighted normalized decision matrix were formed applying the following equations. The values of each normalized term are represented as nij.

$$n_{ij} = \frac{x_{ij}}{\sqrt{\sum_{i=1}^{m} x_{ij}^2}} \text{ ; i= 1, 2, 3, ..., m and j = 1, 2, ..., n} \tag{B.2}$$

$$\tilde{\nabla} = N_D \otimes \tilde{\omega}_j = \begin{bmatrix} \tilde{\nabla}_{1i} & \cdots & \tilde{\nabla}_{1j} & \cdots & \tilde{\nabla}_{1n} \\ \vdots & & \vdots & & \vdots \\ \tilde{\nabla}_{m1} & \cdots & \tilde{\nabla}_{mj} & \cdots & \tilde{\nabla}_{mn} \end{bmatrix} \tag{B.3}$$

The Table B.2 represents the normalized decision matrix.

Table B.2: Normalized decision matrix

Serial No	Name	Density g/cc	Price dollar (2021)	Melting point deg C	Modulus of elasticity GPa
1	Aluminum oxide Al_2O_3	0.1412	0.01944	0.1720	0.1910
2	Barium boride BaB_6	0.1551	0.0648	0.1901	0.1987
3	Chromium carbide Cr_3C_2	0.2378	0.0022	0.1507	0.1961
4	Hafnium Carbide HfC	0.4350	0.3402	0.2511	0.1817
5	Niobium Carbide NbC	0.2711	0.2592	0.2930	0.2322
6	Silicon Carbide SiC	0.1105	0.0146	0.2341	0.2168
7	Tantalum Carbide TaC	0.5099	0.7388	0.3265	0.2838
8	Titanium Carbide, TiC	0.1761	0.0648	0.2565	0.2322

Table B.2: Continued

Serial No	Name	Density g/cc	Price dollar (2021)	Melting point deg C	Modulus of elasticity GPa
9	Vanadium Carbide VC	0.2036	0.1458	0.2285	0.1806
10	Zirconium Carbide ZrC	0.2371	0.2883	0.2846	0.2013
11	Chromium nitride CrN	0.2104	0.0486	0.0879	0.0428
12	CNT	0.0481	0.0065	0.2972	0.4129
13	GNP	0.0806	0.1620	0.2972	0.1088
14	Boron carbide B_4C	0.0899	0.0972	0.1967	0.1495
15	Manganese dioxide MnO_2	0.1792	0.1620	0.0448	0.1426
16	Silicon nitride Si_3N_4	0.1173	0.2754	0.0837	0.1610
17	Aluminum nitride AlN	0.1141	0.0292	0.1423	0.1704
18	Titanium Nitride TiN	0.1861	0.0486	0.2471	0.3097
19	Zirconium Oxide ZrO_2	0.2032	0.0648	0.2302	0.1058
20	Molybdenum Disulfide MoS_2	0.1804	0.0972	0.0992	0.1703

The Table B.3 represents the weighted normalized decision matrix.

Table B.3: Weighted normalized decision matrix

Serial No	Name	Density g/cc	Price dollar (2021)	Melting point deg C	Modulus of elasticity GPa
1	Aluminum oxide Al_2O_3	0.0239	0.01244	0.0169	0.0185
2	Barium boride BaB_6	0.0263	0.04119	0.0186	0.0193
3	Chromium carbide Cr_3C_2	0.0403	0.0014	0.0148	0.0191
4	Hafnium Carbide HfC	0.0737	0.2163	0.0246	0.0177
5	Niobium Carbide NbC	0.0459	0.1648	0.0287	0.0225
6	Silicon Carbide SiC	0.0187	0.0093	0.0229	0.0210
7	Tantalum Carbide TaC	0.0865	0.4696	0.0319	0.0275
8	Titanium Carbide, TiC	0.0299	0.0412	0.0251	0.0225
9	Vanadium Carbide VC	0.0345	0.0927	0.0224	0.0175
10	Zirconium Carbide ZrC	0.0402	0.1833	0.0279	0.0196
11	Chromium nitride CrN	0.0357	0.0309	0.0086	0.0042
12	CNT	0.0082	0.0041	0.0292	0.0401
13	GNP	0.0137	0.1030	0.0292	0.0105
14	Boron carbide B_4C	0.0152	0.0618	0.0193	0.0145
15	Manganese dioxide MnO_2	0.0304	0.1030	0.0044	0.0139

Table B.3: Continued

Serial No	Name	Density g/cc	Price dollar (2021)	Melting point deg C	Modulus of elasticity GPa
16	Silicon nitride Si_3N_4	0.0198	0.1751	0.0082	0.0156
17	Aluminum nitride AlN	0.0193	0.0185	0.0139	0.0165
18	Titanium Nitride TiN	0.0315	0.0309	0.0242	0.0301
19	Zirconium Oxide ZrO_2	0.0344	0.0412	0.0226	0.0103
20	Molybdenum Disulfide MoS_2	0.031	0.0618	0.0097	0.0165

Step 3: Determine the positive and negative ideal solution.

The positive and negative ideal reference solutions were determined using following equations.

$$A^+ = \left\{ \left(\max_j \tilde{V}_{ij}; i \in I \right) \left(\min_j \tilde{V}_{ij}; i \in J \right); i = 1, 2, 3, \ldots, m \right\} \tag{B.4}$$

$$A^- = \left\{ \left(\min_j \tilde{V}_{ij}; i \in I \right) \left(\max_j \tilde{V}_{ij}; i \in J \right); i = 1, 2, 3, \ldots, m \right\} \tag{B.5}$$

Step 4: Determination of performance score.

To determine the performance score, the Euclidean distances from the ideal solution for the alternatives were calculated. The Table B.4 represents the positive and negative ideal solution and performance score.

The Table B.4 represents the ideal solutions and performance score.

Table B.4: Ideal solutions and performance score

Serial No	Name	Positive ideal solution	Negative Ideal solution	Performance score
1	Aluminum oxide Al_2O_3	0.033	0.462	3
2	Barium boride BaB_6	0.050	0.433	8
3	Chromium carbide Cr_3C_2	0.043	0.471	6
4	Hafnium Carbide HfC	0.226	0.255	19
5	Niobium Carbide NbC	0.169	0.309	16
6	Silicon Carbide SiC	0.025	0.466	2
7	Tantalum Carbide TaC	0.475	0.036	20
8	Titanium Carbide, TiC	0.049	0.433	7
9	Vanadium Carbide VC	0.098	0.381	13
10	Zirconium Carbide ZrC	0.186	0.291	18
11	Chromium nitride CrN	0.059	0.442	10
12	CNT	0.004	0.474	1
13	GNP	0.106	0.375	14
14	Boron carbide B_4C	0.067	0.414	11
15	Manganese dioxide MnO_2	0.111	0.371	15
16	Silicon nitride Si_3N_4	0.177	0.302	17

Table B.4: Continued

Serial No	Name	Positive ideal solution	Negative Ideal solution	Performance score
17	Aluminum nitride AlN	0.036	0.456	4
18	Titanium Nitride TiN	0.0397	0.443	5
19	Zirconium Oxide ZrO_2	0.057	0.432	9
20	Molybdenum Disulfide MoS_2	0.072	0.412	12

Appendix – C: Reinforcement Selection Using Simple Additive Weighting (SAW) Method

This study examines methods for assessing reinforcements performance. The performance scores different reinforcements were evaluated using fuzzy simple additive weighting (SAW) approach. Performance score is the reinforcements reaction variable to its attributes. The following stages were used in the process.

Step 1: Developing the decision matrix

The decision matrix for the SAW analysis was formed applying the following equation, the alternatives or the variants of reinforcements are represented as A1, A2, …, Am. The criteria such as density, price, melting point, modulus of elasticity are represented as C1, C2, Cm. The values of each criterion are represented as Xij.

$$D = \begin{matrix} \\ A_1 \\ A_2 \\ \vdots \\ A_m \end{matrix} \begin{bmatrix} C_1 & C_2 & \cdots & C_n \\ x_{11} & x_{12} & \cdots & x_{1n} \\ x_{21} & x_{22} & \cdots & x_{2n} \\ \vdots & \vdots & \vdots & \vdots \\ x_{m1} & x_{m2} & \cdots & x_{mn} \end{bmatrix} \quad (C.1)$$

The Table C.1 represents the decision matrix.

Table C.1: Decision matrix for alternative particulate reinforcements

Serial No	Name	Density g/cc	Price dollar (2021)	Melting point deg C	Modulus of elasticity GPa
1	Aluminum oxide Al_2O_3	3.96	6	2054	370
2	Barium boride BaB_6	4.35	20	2270	385
3	Chromium carbide Cr_3C_2	6.67	0.66	1800	380

Table C.1: Continued

Serial No	Name	Density g/cc	Price dollar (2021)	Melting point deg C	Modulus of elasticity GPa
4	Hafnium Carbide HfC	12.2	105	3000	352
5	Niobium Carbide NbC	7.6	80	3500	450
6	Silicon Carbide SiC	3.1	4.5	2797	420
7	Tantalum Carbide TaC	14.3	228	3900	550
8	Titanium Carbide, TiC	4.94	20	3065	450
9	Vanadium Carbide VC	5.71	45	2730	350
10	Zirconium Carbide ZrC	6.65	89	3400	390
11	Chromium nitride CrN	5.9	15	1050	83
12	CNT	1.35	2	3550	800
13	GNP	2.26	50	3550	210
14	Boron carbide B_4C	2.52	30	2350	289.65
15	Manganese dioxide MnO_2	5.026	50	535	276.34
16	Silicon nitride Si_3N_4	3.29	85	1000	310
17	Aluminum nitride AlN	3.2	9	1700	330

Table C.1: Continued

Serial No	Name	Density g/cc	Price dollar (2021)	Melting point deg C	Modulus of elasticity GPa
18	Titanium Nitride TiN	5.22	15	2950	600
19	Zirconium Oxide ZrO_2	5.7	20	2750	205
20	Molybdenum Disulfide MoS_2	5.06	30	1185	330

Step 2: Formulate the normalized decision matrix B:

The normalized decision matrix for the SAW method and weighted normalized decision matrix were formed applying the following equations. The values of each normalized term are represented as nij.

$$B = [b_{ij}] \tag{C.2}$$

$$b_{ij} = \begin{cases} \dfrac{a_{ij}}{\max\limits_i a_{ij}} & if\ j \epsilon C_p \\ \dfrac{\max\limits_i a_{ij}}{a_{ij}} & if\ j \epsilon C_n \end{cases} \tag{C.3}$$

The table C.2 represents the normalized decision matrix.

Table C.2: Normalized decision matrix

Serial No	Name	Density g/cc	Price dollar (2021)	Melting point deg C	Modulus of elasticity GPa
1	Aluminum oxide Al_2O_3	0.03633	0.0066	0.0418	0.04913
2	Barium boride BaB_6	0.0399	0.022	0.0462	0.0511

Table C.2: Continued

Serial No	Name	Density g/cc	Price dollar (2021)	Melting point deg C	Modulus of elasticity GPa
3	Chromium carbide Cr_3C_2	0.0612	0.00073	0.0366	0.0505
4	Hafnium Carbide HfC	0.1119	0.1161	0.0611	0.047
5	Niobium Carbide NbC	0.0697	0.0885	0.0712	0.0598
6	Silicon Carbide SiC	0.0284	0.00498	0.0569	0.0558
7	Tantalum Carbide TaC	0.1312	0.2522	0.0794	0.0730
8	Titanium Carbide, TiC	0.0453	0.02212	0.0624	0.0598
9	Vanadium Carbide VC	0.0524	0.0498	0.0556	0.0464
10	Zirconium Carbide ZrC	0.0610	0.0984	0.0692	0.0518
11	Chromium nitride CrN	0.0542	0.0166	0.0214	0.0110
12	CNT	0.01239	0.0022	0.07225	0.10623
13	GNP	0.02073	0.0553	0.07225	0.0279
14	Boron carbide B_4C	0.02312	0.03318	0.0478	0.0385
15	Manganese dioxide MnO_2	0.04611	0.0553	0.0109	0.0367
16	Silicon nitride Si_3N_4	0.03019	0.0940	0.0204	0.04116
17	Aluminum nitride AlN	0.02936	0.0099	0.0346	0.0438
18	Titanium Nitride TiN	0.0478	0.0166	0.0604	0.0797

Table C.2: Continued

Serial No	Name	Density g/cc	Price dollar (2021)	Melting point deg C	Modulus of elasticity GPa
19	Zirconium Oxide ZrO_2	0.0523	0.0221	0.056	0.027
20	Molybdenum Disulfide MoS_2	0.0464	0.0332	0.024	0.0438

Step 3: The weights assigned normalized decision matrix to evaluate the weights of the selected attributes using the following equations:

$$y_{ij} = b_{ij}.w_j \tag{C.4}$$

$$Y_i = \sum_{j=1}^{n} y_{ij} \tag{C.5}$$

The Table C.3 represents the weighted normalized decision matrix.

Table C.3: Weighted normalized decision matrix

Serial No	Name	Density g/cc	Price dollar (2021)	Melting point deg C	Modulus of elasticity GPa
1	Aluminum oxide Al_2O_3	0.341	0.11	0.527	0.4625
2	Barium boride BaB_6	0.311	0.033	0.5821	0.4813
3	Chromium carbide Cr_3C_2	0.2024	1	0.4616	0.475
4	Hafnium Carbide HfC	0.1107	0.0063	0.7692	0.44
5	Niobium Carbide NbC	0.1776	0.00825	0.8974	0.5625
6	Silicon Carbide SiC	0.4355	0.14667	0.7172	0.525

Table C.3: Continued

Serial No	Name	Density g/cc	Price dollar (2021)	Melting point deg C	Modulus of elasticity GPa
7	Tantalum Carbide TaC	0.0944	0.0029	1	0.6875
8	Titanium Carbide, TiC	0.2734	0.033	0.786	0.5625
9	Vanadium Carbide VC	0.2364	0.0147	0.7	0.4375
10	Zirconium Carbide ZrC	0.2031	0.007	0.8718	0.4875
11	Chromium nitride CrN	0.2289	0.044	0.269	0.1038
12	CNT	1	0.33	0.911	1
13	GNP	0.597	0.0132	0.911	0.2625
14	Boron carbide B_4C	0.536	0.022	0.6026	0.3621
15	Manganese dioxide MnO_2	0.269	0.0132	0.1373	0.345
16	Silicon nitride Si_3N_4	0.4103	0.0078	0.256	0.388
17	Aluminum nitride AlN	0.422	0.0733	0.4359	0.4125
18	Titanium Nitride TiN	0.2586	0.044	0.7564	0.75
19	Zirconium Oxide ZrO_2	0.2368	0.033	0.70513	0.256
20	Molybdenum Disulfide MoS_2	0.2668	0.022	0.3038	0.4125

Step 4: Rank the alternatives

The assessment score of the chosen attributes is utilized to evaluate the ranking of the alternatives given in Table 3.8.

The Table C.4 represents the weight of alternatives and rank.

Table C.3: Weight of alternatives and rank

Serial No	Name	Y_i	Rank
1	Aluminum oxide Al_2O_3	0.2252	5
2	Barium boride BaB_6	0.1788	12
3	Chromium carbide Cr_3C_2	0.7607	1
4	Hafnium Carbide HfC	0.1439	16
5	Niobium Carbide NbC	0.1813	11
6	Silicon Carbide SiC	0.2901	3
7	Tantalum Carbide TaC	0.1863	10
8	Titanium Carbide, TiC	0.2014	8
9	Vanadium Carbide VC	0.1629	14
10	Zirconium Carbide ZrC	0.1753	13
11	Chromium nitride CrN	0.1041	19
12	CNT	0.5658	2
13	GNP	0.2280	4
14	Boron carbide B_4C	0.2006	9

Table C.3 (Continued)

Serial No	Name	Y_i	Rank
15	Manganese dioxide MnO_2	0.1002	20
16	Silicon nitride Si_3N_4	0.1369	17
17	Aluminum nitride AlN	0.2014	7
18	Titanium Nitride TiN	0.2205	6
19	Zirconium Oxide ZrO_2	0.1581	15
20	Molybdenum Disulfide MoS_2	0.1291	18

Appendix – D: Reinforcement Selection Using Multi-Objective Optimization by Ratio Analysis (MOORA) Method

Introduced by Brauers and Zavadskas as a multi-criteria decision making (MCDM), MOORA method technique is used for solving complex multi-objective problems in transition economics. The following stages were used in the process.

Step 1: Developing the decision matrix

The decision matrix for the MOORA analysis was formed applying the following equation, the alternatives or the variants of reinforcements are represented as A1, A2, …, Am. The criteria such as density, price, melting point, modulus of elasticity are represented as C1, C2, Cm. The values of each criterion are represented as Xij.

$$D = \begin{matrix} & \\ A_1 \\ A_2 \\ \vdots \\ A_m \end{matrix} \begin{bmatrix} C_1 & C_2 & \cdots & C_n \\ x_{11} & x_{12} & \cdots & x_{1n} \\ x_{21} & x_{22} & \cdots & x_{2n} \\ \vdots & \vdots & \vdots & \vdots \\ x_{m1} & x_{m2} & \cdots & x_{mn} \end{bmatrix} \tag{D.1}$$

The Table D.1 represents the decision matrix.

Table D.1: Decision matrix for alternative particulate reinforcements

Serial No	Name	Density g/cc	Price dollar (2021)	Melting point deg C	Modulus of elasticity GPa
1	Aluminum oxide Al_2O_3	3.96	6	2054	370
2	Barium boride BaB_6	4.35	20	2270	385
3	Chromium carbide Cr_3C_2	6.67	0.66	1800	380
4	Hafnium Carbide HfC	12.2	105	3000	352
5	Niobium Carbide NbC	7.6	80	3500	450
6	Silicon Carbide SiC	3.1	4.5	2797	420

Table D.1: Continued

Serial No	Name	Density g/cc	Price dollar (2021)	Melting point deg C	Modulus of elasticity GPa
7	Tantalum Carbide TaC	14.3	228	3900	550
8	Titanium Carbide, TiC	4.94	20	3065	450
9	Vanadium Carbide VC	5.71	45	2730	350
10	Zirconium Carbide ZrC	6.65	89	3400	390
11	Chromium nitride CrN	5.9	15	1050	83
12	CNT	1.35	2	3550	800
13	GNP	2.26	50	3550	210
14	Boron carbide B_4C	2.52	30	2350	289.65
15	Manganese dioxide MnO_2	5.026	50	535	276.34
16	Silicon nitride Si_3N_4	3.29	85	1000	310
17	Aluminum nitride AlN	3.2	9	1700	330
18	Titanium Nitride TiN	5.22	15	2950	600
19	Zirconium Oxide ZrO_2	5.7	20	2750	205
20	Molybdenum Disulfide MoS_2	5.06	30	1185	330

Step 2: Developing the normalized and fuzzy weighted normalized decision matrix.

This step employs ratio scheme in which each alternative performance on criterion is equated to denominator representing all alternatives concerning that criterion.

$$D_n = \frac{x_{ij}}{\sqrt{\sum_{j=1}^{v} x_{ij}^2}} \tag{D.2}$$

Where, D_n is denoted as dimensionless number having interval of [0,1] showing the normalized performances measures of jth alternative related to ith objective. The Table D.2 represents the normalized decision matrix.

Table D.2: Normalized decision matrix

Serial No	Name	Density g/cc	Price dollar (2021)	Melting point deg C	Modulus of elasticity GPa
1	Aluminum oxide Al_2O_3	0.1417	0.0194	0.1742	0.2043
2	Barium boride BaB_6	0.1557	0.0648	0.1926	0.2126
3	Chromium carbide Cr_3C_2	0.2387	0.0021	0.1527	0.2098
4	Hafnium Carbide HfC	0.4366	0.3404	0.2545	0.1944
5	Niobium Carbide NbC	0.2720	0.2593	0.2969	0.2485
6	Silicon Carbide SiC	0.1109	0.0146	0.2373	0.2319
7	Tantalum Carbide TaC	0.5118	0.7391	0.3308	0.3037
8	Titanium Carbide, TiC	0.1768	0.0648	0.2600	0.2485
9	Vanadium Carbide VC	0.2043	0.1459	0.2316	0.1933
10	Zirconium Carbide ZrC	0.2380	0.2885	0.2884	0.2154

Table D.2: Continued

Serial No	Name	Density g/cc	Price dollar (2021)	Melting point deg C	Modulus of elasticity GPa
11	Chromium nitride CrN	0.2111	0.0486	0.0891	0.0458
12	CNT	0.0483	0.0065	0.3011	0.4418
13	GNP	0.0809	0.1621	0.3011	0.1160
14	Boron carbide B_4C	0.0902	0.0972	0.1993	0.1600
15	Manganese dioxide MnO_2	0.1799	0.1621	0.0454	0.1526
16	Silicon nitride Si_3N_4	0.1177	0.2755	0.0848	0.1712
17	Aluminum nitride AlN	0.1145	0.0292	0.1442	0.1822
18	Titanium Nitride TiN	0.1868	0.0486	0.2502	0.3313
19	Zirconium Oxide ZrO_2	0.2040	0.0648	0.2333	0.1132
20	Molybdenum Disulfide MoS_2	0.1811	0.0972	0.1005	0.1822

Step 3: Formulation of optimization problem by adding normalized performance measures.

The maximization (beneficiary objective – benefits) and subtracted for minimization (non-beneficiary objective – cost) solutions were determined using following equations.

$$N_j = \sum_{i=1}^{i=m} D_n - \sum_{i=m+1}^{u} D_n \tag{D.3}$$

Where, the maximization objective/criteria numbers are denoted as m and (u-m) denotes the number of objectives to be minimized. The term N_j represents normalized evaluation of alternative j with respect to all criteria.

Step 4: Determination of weightage to one objective. The following equation can be used to determine weightage normalized decision matrix.

$$N_j = \sum_{i=1}^{i=m} w_i * D_n - \sum_{i=m+1}^{u} w_i * D_n \qquad \text{(D.4)}$$

The table D.3 represents the weighted normalized decision matrix.

Table D.3: Weighted normalized decision matrix

Serial No	Name	Density g/cc	Price dollar (2021)	Melting point deg C	Modulus of elasticity GPa
1	Aluminum oxide Al_2O_3	0.0238	0.0123	0.0181	0.0192
2	Barium boride BaB_6	0.0261	0.0411	0.0200	0.0200
3	Chromium carbide Cr_3C_2	0.0401	0.0014	0.0159	0.0197
4	Hafnium Carbide HfC	0.0733	0.2158	0.0265	0.0183
5	Niobium Carbide NbC	0.0457	0.1644	0.0309	0.0233
6	Silicon Carbide SiC	0.0186	0.0092	0.0247	0.0218
7	Tantalum Carbide TaC	0.0859	0.4686	0.0344	0.0285
8	Titanium Carbide, TiC	0.0297	0.0411	0.0270	0.0233
9	Vanadium Carbide VC	0.0343	0.0925	0.0241	0.0182
10	Zirconium Carbide ZrC	0.0400	0.1829	0.0300	0.0202
11	Chromium nitride CrN	0.0355	0.0308	0.0093	0.0043
12	CNT	0.0081	0.0041	0.0313	0.0415
13	GNP	0.0136	0.1028	0.0313	0.0109

Table D.3: Continued

Serial No	Name	Density g/cc	Price dollar (2021)	Melting point deg C	Modulus of elasticity GPa
14	Boron carbide B_4C	0.0151	0.0617	0.0207	0.0150
15	Manganese dioxide MnO_2	0.0302	0.1028	0.0047	0.0143
16	Silicon nitride Si_3N_4	0.0198	0.1747	0.0088	0.0161
17	Aluminum nitride AlN	0.0192	0.0185	0.0150	0.0171
18	Titanium Nitride TiN	0.0314	0.0308	0.0260	0.0311
19	Zirconium Oxide ZrO_2	0.0343	0.0411	0.0243	0.0106
20	Molybdenum Disulfide MoS_2	0.0304	0.0617	0.0105	0.0171

The Table D.4 represents the sort and rank the results of available alternative depending on the normalized performance of N_j values.

Table D.4: Performance score

Serial No	Name	Normalized performance measure (N_j)	Performance score
1	Aluminum oxide Al_2O_3	0.0012	3
2	Barium boride BaB_6	-0.0272	8
3	Chromium carbide Cr_3C_2	-0.0058	6
4	Hafnium Carbide HfC	-0.2444	19

Table D.4: Continued

Serial No	Name	Normalized performance measure (N_j)	Performance score
5	Niobium Carbide NbC	-0.1559	16
6	Silicon Carbide SiC	0.0186	2
7	Tantalum Carbide TaC	-0.4916	20
8	Titanium Carbide, TiC	-0.0204	7
9	Vanadium Carbide VC	-0.0846	14
10	Zirconium Carbide ZrC	-0.1727	18
11	Chromium nitride CrN	-0.0527	11
12	CNT	0.0606	1
13	GNP	-0.0741	13
14	Boron carbide B_4C	-0.0410	10
15	Manganese dioxide MnO_2	-0.1139	15
16	Silicon nitride Si_3N_4	-0.1696	17
17	Aluminum nitride AlN	-0.0056	5

Table D.4: Continued

Serial No	Name	Normalized performance measure (N_j)	Performance score
18	Titanium Nitride TiN	-0.0050	4
19	Zirconium Oxide ZrO_2	-0.0405	9
20	Molybdenum Disulfide MoS_2	-0.0645	12

Appendix – E: Sensitivity Analysis by Inducing Disturbance on Fuzzy Criteria Weights

To carry out the sensitivity analysis of the TOPSIS, MOORA, and SAW method, an investigation was conducted varying the weightage ratio of the fuzzy criteria weights. Table E.1 represents an example of the change in the relative fuzzy criteria weights.

Table E.1 Sensitivity analysis weight factor distribution

Serial no	Weightage ratio	Beneficiary factor weightage	Non-beneficiary factor weightage
1	Entropy weightage		
2	50% weightage	0.5	0.2
3	Beneficiary 60%, Non-beneficiary 40%	0.6	0.4
4	Beneficiary 70%, Non-beneficiary 30%	0.7	0.3
5	Beneficiary 80%, Non-beneficiary 20%	0.8	0.2
6	Beneficiary 90%, Non-beneficiary 10%	0.9	0.1
7	Beneficiary 10%, Non-beneficiary 90%	0.1	0.9
8	Beneficiary 20%, Non-beneficiary 80%	0.2	0.8
9	Beneficiary 30%, Non-beneficiary 70%	0.3	0.7
10	Beneficiary 40%, Non-beneficiary 60%	0.4	0.6

Following the Table E.1, sensitivity study was performed. Fig E.1 to E.4 shows the change in the ranking derived from TOPSIS, MOORA and SAW method due to sequential induced disturbance on the criteria weights.

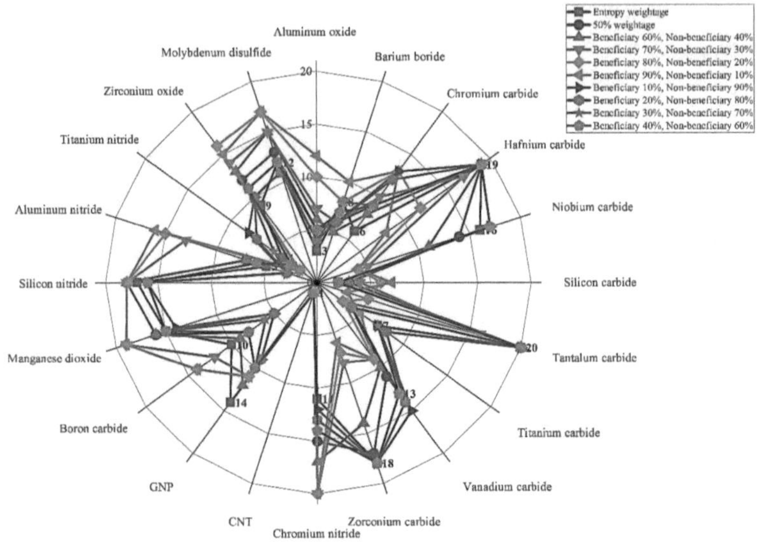

Fig. E.1: Ranking of the particulate reinforcements derived from TOPSIS analysis with induced disturbance on the weightages.

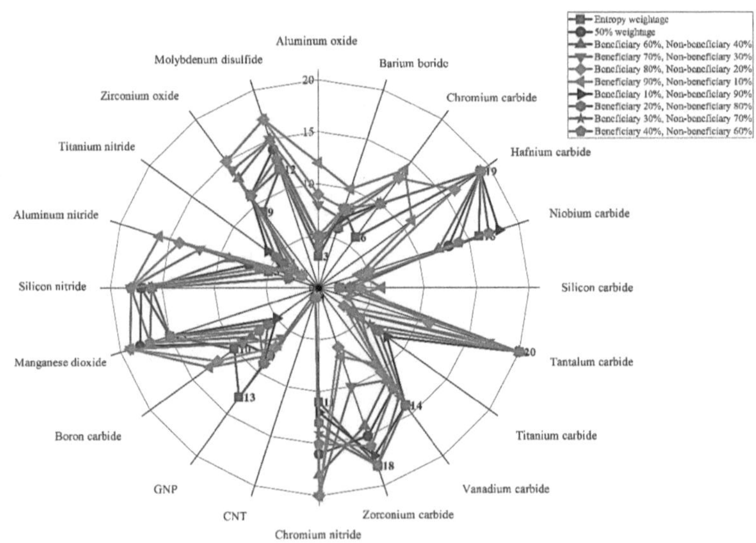

Fig. E.2: Ranking of the particulate reinforcements derived from MOORA analysis with induced disturbance on the weightages.

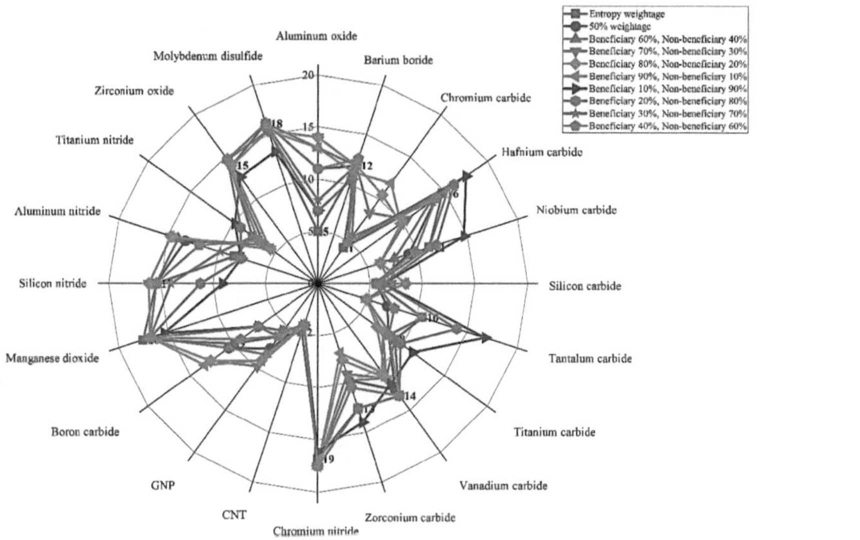

Fig. E.3: Ranking of the particulate reinforcements derived from SAW analysis with induced disturbance on the weightages.

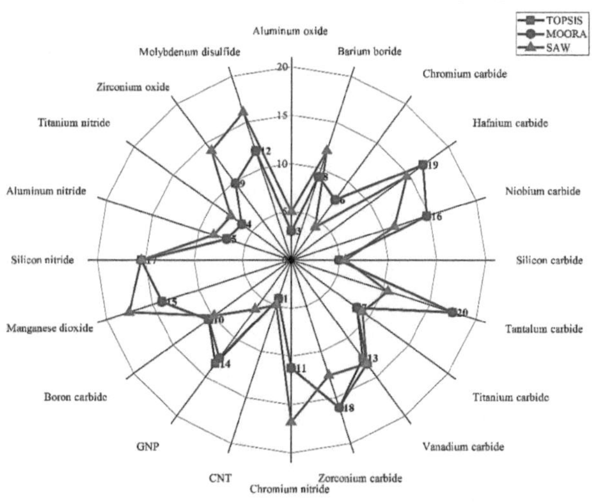

Fig. E.4: Ranking of the particulate reinforcements derived from TOPSIS, MOORA and SAW analysis with entropy weightages.

From the figures it can be observed, the non-beneficiary factors found to be sensitive to the performance ranking. No variation for CNT, silicon carbide and alumina was observed in TOPSIS and MOORA analysis. Minor variations found for the GNP and Titanium nitride. The rest of the reinforcements were largely sensitive to the variations of the weightage.

Appendix – F: Optimization of Drilling Parameters Using Grey Relational Analysis

The factors and their levels considered in this study are shown in Table E.1. Experiments are conducted with two factors each at three levels and hence a three level orthogonal array (OA) is chosen.

Table F.1 Factors and Levels for Grey Relational Analysis

Parameters	Unit	Level		
		1	2	3
Cutting Speed (V)	m/min	400	1140	2260
Feed (F)	mm/rev	0.1	0.15	0.2

Step 1: Calculate S/N ratio for the corresponding responses using the following formula.

Larger - the – better

$$\text{S/N ratio}(\eta) = -10\log_{10}\left(\frac{1}{n}\right)\sum_{i=1}^{n}\frac{1}{y_{ij}^2} \qquad \text{(F.1)}$$

where n=number of replications yij=observed response value where i=1, 2, ... n; j=1, 2...k This is applied for problem where maximization of the quality characteristic of interest is sought. This is referred as the larger-the-better type problem.

Smaller - the – better

$$\text{S/N ratio}(\eta) = -10\log_{10}\left(\frac{1}{n}\sum_{i=1}^{n}y_{ij}^2\right) \qquad \text{(F.2)}$$

This is termed as the smaller-the-better type problem where minimization of the characteristic is intended.

Nominal - the – best

$$\text{S/N ratio } (\eta) = 10 \log_{10}\left(\frac{\mu^2}{\sigma^2}\right)$$

$$\text{Where } \mu = \frac{y1+y2+y3+.......+yn}{n} \qquad \text{(F.3)}$$

$$\sigma^2 = \frac{\sum (yi-\bar{y})^2}{n-1}$$

This is called nominal-the-best type of problem where one tries to minimize the mean squared error around a specific target value. Adjusting the mean on target by any means renders the problem to a constrained optimization problem.

Normalization is a transformation performed on a single data input to distribute the data evenly and scale it into an acceptable range for further analysis.

The following tables shows the matrix for developed composites.

Table F.2: Orthogonal array with factors and responses for Al-alloy

Specimen 1- Al-alloy					
RPM	Feed rate, mm/rev	In diameter (Top), mm	Out diameter (Bottom), (mm)	Taper angle	Delamination
	0.1	8.5	7.9	0.0299	1.2613
400	0.15	8.5	8	0.02499	1.2611
	0.2	8.45	8	0.0225	1.2537
	0.1	8.2	8	0.0099	1.2166
1140	0.15	8.2	8	0.0099	1.2166
	0.2	8.1	8	0.0049	1.2018
	0.1	8.1	7.9	0.0099	1.2018
2260	0.15	8.1	7.9	0.0099	1.2018
	0.2	8.1	8	0.0049	1.2018

Table F.3: Orthogonal array with factors and responses for Al-alloy and CNT composites

Specimen 2- Al-alloy + CNT					
RPM	Feed rate, mm/rev	In diameter (Top), mm	Out diameter (Bottom), (mm)	Taper angle	Delamination
400	0.1	8.3	7.9	0.0199	1.2315
	0.15	8.3	8	0.0149	1.2315
	0.2	8.2	7.85	0.0175	1.2166
1140	0.1	8.1	7.9	0.0099	1.2018
	0.15	8.1	7.9	0.0099	1.2018
	0.2	8.2	8	0.0099	1.2166
2260	0.1	8.2	8.1	0.0049	1.2166
	0.15	8.1	8	0.0049	1.2018
	0.2	8.15	8	0.0075	1.2092

Table F.4: Orthogonal array with factors and responses for Al-alloy, SiC and CNT composites

Specimen 3- Al-alloy + CNT +SiC					
RPM	Feed rate, mm/rev	In diameter (Top), mm	Out diameter (Bottom), (mm)	Taper angle	Delamination
400	0.1	8.5	7.9	0.0299	1.2611
	0.15	8.2	8	0.0099	1.2166
	0.2	8.3	8.1	0.0099	1.2315
1140	0.1	8.1	7.9	0.0099	1.2018
	0.15	8.2	7.9	0.0149	1.2018
	0.2	8.1	7.9	0.0099	1.2018
2260	0.1	8.1	8	0.0049	1.2018
	0.15	8.1	7.9	0.0099	1.2018
	0.2	8.2	8	0.0099	1.2166

Table F.5: Orthogonal array with factors and responses for Al-alloy, SiC, alumina and CNT composites

Specimen 4- Al + CNT + SiC + Al2O3					
RPM	Feed rate, mm/rev	In diameter (Top), mm	Out diameter (Bottom), (mm)	Taper angle	Delamination
400	0.1	8.3	8	0.0149	1.2315
	0.15	8.2	7.9	0.0149	1.2167
	0.2	8.1	8	0.0049	1.2018
1140	0.1	8.2	8	0.0099	1.2167
	0.15	8.2	7.9	0.0149	1.2166
	0.2	8.1	7.9	0.0099	1.2018
2260	0.1	8.1	8	0.0049	1.2018
	0.15	8.15	8	0.0075	1.2092
	0.2	8	7.9	0.0049	1.1869

The S/N ratios for the given responses are represented in the following tables.

Table F.6: S/N ratio values for aluminum alloy

Specimen 1- Al -alloy					
RPM	Feed rate, mm/rev	In diameter (Top), mm	Out diameter (Bottom), (mm)	Taper angle	Delamination
400	0.1	-18.5884	-17.9525	30.4602	-2.0152
	0.15	-18.5884	-18.0618	32.0430	-2.0152
	0.2	-18.5371	-18.0618	32.9578	-1.9639
1140	0.1	-18.2763	-18.0618	40.0003	-1.7031
	0.15	-18.2763	-18.0618	40.0003	-1.7031
	0.2	-18.1697	-18.0618	46.0207	-1.5965
2260	0.1	-18.1697	-17.9525	40.0003	-1.5965
	0.15	-18.1697	-17.9525	40.0003	-1.5965
	0.2	-18.1697	-18.0618	46.0207	-1.5965

Table F.7: S/N ratio values for aluminum alloy and CNT composites

Specimen 2- Al –alloy + CNT					
RPM	Feed rate, mm/rev	In diameter (Top), mm	Out diameter (Bottom), (mm)	Taper angle	Delamination
400	0.1	-18.3816	-17.9525	33.9806	-1.8084
	0.15	-18.3816	-18.0618	36.4788	-1.8084
	0.2	-18.2763	-17.8974	35.1401	-1.7031
1140	0.1	-18.1697	-17.9525	40.0003	-1.5965
	0.15	-18.1697	-17.9525	40.0003	-1.5965
	0.2	-18.2763	-18.0618	40.0003	-1.7031
2260	0.1	-18.2763	-18.1697	46.0207	-1.7031
	0.15	-18.1697	-18.0618	46.0207	-1.5965
	0.2	-18.2232	-18.0618	42.4989	-1.6500

Table F.8: S/N ratio values for aluminum alloy, CNT and silicon carbide composites

Specimen 3- Al –alloy + CNT+ Silicon carbide					
RPM	Feed rate, mm/rev	In diameter (Top), mm	Out diameter (Bottom), (mm)	Taper angle	Delamination
400	0.1	-18.5884	-17.9525	30.4602	-2.0152
	0.15	-18.2763	-18.0618	40.0003	-1.7031
	0.2	-18.3816	-18.1697	40.0003	-1.8084
1140	0.1	-18.1697	-17.9525	40.0003	-1.5965
	0.15	-18.2763	-17.9525	36.4788	-1.7031
	0.2	-18.1697	-17.9525	40.0003	-1.5965
2260	0.1	-18.1697	-18.0618	46.0207	-1.5965
	0.15	-18.1697	-17.9525	40.0003	-1.5965
	0.2	-18.2763	-18.0618	40.0003	-1.7031

Table F.9: S/N ratio values for aluminum alloy, silicon carbide, alumina and CNT composites

Specimen 4- Al –alloy + CNT+ Silicon carbide + Alumina					
RPM	Feed rate, mm/rev	In diameter (Top), mm	Out diameter (Bottom), (mm)	Taper angle	Delamination
400	0.1	-18.3816	-18.0618	36.4788	-1.8084
	0.15	-18.2763	-17.9525	36.4788	-1.7031
	0.2	-18.1697	-18.0618	46.0207	-1.5965
1140	0.1	-18.2763	-18.0618	40.0003	-1.7031
	0.15	-18.2763	-18.0618	40.0003	-1.7031
	0.2	-18.1697	-17.9525	40.0003	-1.5965
2260	0.1	-18.1697	-18.0618	46.0207	-1.5965
	0.15	-18.2232	-18.0618	42.4989	-1.6500
	0.2	-18.0618	-17.9525	46.0207	-1.4886

Step 2: Normalize the S/N ratio values

The following equations is used to normalized the obtained signal to noise values.

$$Z_{ij} = \frac{y_{ij} - \min(y_{ij}, i = 1, 2, ...n)}{\max(y_{ij}, i = 1, 2, ...n) - \min(y_{ij}, i = 1, 2, ...n)} \quad (F.4)$$

$$Z_{ij} = \frac{\max(y_{ij}, i = 1, 2, ...n) - y_{ij}}{\max(y_{ij}, i = 1, 2, ...n) - \min(y_{ij}, i = 1, 2, ...n)} \quad (F.5)$$

$$Z_{ij} = \frac{(y_{ij} - \text{Target}) - \min(|y_{ij} - \text{Target}|, i = 1, 2, ...n)}{\max(|y_{ij} - \text{Target}|, i = 1, 2, ...n) - \min(|y_{ij} - \text{Target}|, i = 1, 2, ...n)} \quad (F.6)$$

Table F.10: Normalized S/N ratio values for aluminum alloy

Specimen 1- Al –alloy					
RPM	Feed rate, mm/rev	In diameter (Top), mm	Out diameter (Bottom), (mm)	Taper angle	Delamination
	0.1	1.0000	0.0000	1.0000	1.0000
400	0.15	1.0000	1.0000	0.8983	1.0000
	0.2	0.8776	1.0000	0.8395	0.8776
	0.1	0.2546	1.0000	0.3869	0.2546
1140	0.15	0.2546	1.0000	0.3869	0.2546
	0.2	0.0000	1.0000	0.0000	0.0000
	0.1	0.0000	0.0000	0.3869	0.0000
2260	0.15	0.0000	0.0000	0.3869	0.0000
	0.2	0.0000	1.0000	0.0000	0.0000

Table F.11: Normalized S/N ratio values for aluminum alloy and CNT composite

Specimen 2- Al –alloy + CNT					
RPM	Feed rate, mm/rev	In diameter (Top), mm	Out diameter (Bottom), (mm)	Taper angle	Delamination
	0.1	1.0000	-1.0250	-246.1526	-77.2266
400	0.15	1.0000	-0.5093	-257.9446	-77.2266
	0.2	0.5030	-1.2853	-251.6259	-77.7235
	0.1	0.0000	-1.0250	-274.5662	-78.2266
1140	0.15	0.0000	-1.0250	-274.5662	-78.2266
	0.2	0.5030	-0.5093	-274.5662	-77.7235
	0.1	0.5030	0.0000	-302.9828	-77.7235
2260	0.15	0.0000	-0.5093	-302.9828	-78.2266
	0.2	0.2523	-0.5093	-286.3599	-77.9743

Table F.12: Normalized S/N ratio values for aluminum alloy, silicon carbide and CNT composite

Specimen 3- Al –alloy + CNT + Silicon carbide					
RPM	Feed rate, mm/rev	In diameter (Top), mm	Out diameter (Bottom), (mm)	Taper angle	Delamination
	0.1	1.0000	-0.5187	-116.1510	-38.5846
400	0.15	0.2546	-0.2577	-138.9373	-39.3300
	0.2	0.5060	0.0000	-138.9373	-39.0786
	0.1	0.0000	-0.5187	-138.9373	-39.5846
1140	0.15	0.2546	-0.5187	-130.5263	-39.3300
	0.2	0.0000	-0.5187	-138.9373	-39.5846
	0.1	0.0000	-0.2577	-153.3168	-39.5846
2260	0.15	0.0000	-0.5187	-138.9373	-39.5846
	0.2	0.2546	-0.2577	-138.9373	-39.3300

Table F.13: Normalized S/N ratio values for aluminum alloy, silicon carbide and CNT composite

Specimen 4- Al –alloy + CNT + Silicon carbide + Alumina					
RPM	Feed rate, mm/rev	In diameter (Top), mm	Out diameter (Bottom), (mm)	Taper angle	Delamination
	0.1	1.0000	0.0000	-170.5663	-50.8298
400	0.15	0.6707	-0.3417	-170.5663	-51.1590
	0.2	0.3374	0.0000	-200.4067	-51.4923
	0.1	0.6707	0.0000	-181.5790	-51.1590
1140	0.15	0.6707	0.0000	-181.5790	-51.1590
	0.2	0.3374	-0.3417	-181.5790	-51.4923
	0.1	0.3374	0.0000	-200.4067	-51.4923
2260	0.15	0.5046	0.0000	-189.3931	-51.3252
	0.2	0.0000	-0.3417	-200.4067	-51.8298

Step 3: Perform the Grey relational analysis

Calculate the grey relational co-efficient for the normalized S/N ratio value has been calculated using the following equation.

$$\gamma(y_o(k), y_i(k)) = \frac{\Delta \min + \xi \Delta \max}{\Delta_{oj}(k) + \xi \Delta \max} \quad \text{(F.7)}$$

Table F.14: Grey relational co-efficient for aluminum alloy

Specimen 1- Al –alloy					
RPM	Feed rate, mm/rev	In diameter (Top), mm	Out diameter (Bottom), (mm)	Taper angle	Delamination
	0.1	0.3333	1.0000	0.3333	0.3333
400	0.15	0.3333	0.3333	0.3576	0.3333
	0.2	0.3629	0.3333	0.3733	0.3629
	0.1	0.6626	0.3333	0.5638	0.6626
1160	0.15	0.6626	0.3333	0.5638	0.6626
	0.2	1.0000	0.3333	1.0000	1.0000
	0.1	1.0000	1.0000	0.5638	1.0000
2260	0.15	1.0000	1.0000	0.5638	1.0000
	0.2	1.0000	0.3333	1.0000	1.0000

Table F.15: Grey relational co-efficient for aluminum alloy and CNT

Specimen 2- Al –alloy + CNT					
RPM	Feed rate, mm/rev	In diameter (Top), mm	Out diameter (Bottom), (mm)	Taper angle	Delamination
	0.1	0.3333	-0.9524	-0.0020	-0.0065
400	0.15	0.3333	-53.7746	-0.0019	-0.0065
	0.2	0.4985	-0.6367	-0.0020	-0.0065
	0.1	1.0000	-0.9524	-0.0018	-0.0064
1140	0.15	1.0000	-0.9524	-0.0018	-0.0064
	0.2	0.4985	-53.7746	-0.0018	-0.0065
2260	0.1	0.4985	1.0000	-0.0017	-0.0065
	0.15	1.0000	-53.7746	-0.0017	-0.0064

Specimen 2- Al –alloy + CNT					
	0.2	0.6646	-53.7746	-0.0017	-0.0065

Table F.16: Grey relational co-efficient for aluminum alloy, silicon carbide and CNT

Specimen 3- Al –alloy + CNT + Silicon carbide					
RPM	Feed rate, mm/rev	In diameter (Top), mm	Out diameter (Bottom), (mm)	Taper angle	Delamination
400	0.1	0.3333	-26.7715	-0.0043	-0.0131
	0.15	0.6626	2.0637	-0.0036	-0.0129
	0.2	0.4970	1.0000	-0.0036	-0.0130
1140	0.1	1.0000	-26.7715	-0.0036	-0.0128
	0.15	0.6626	-26.7715	-0.0038	-0.0129
	0.2	1.0000	-26.7715	-0.0036	-0.0128
2260	0.1	1.0000	2.0637	-0.0033	-0.0128
	0.15	1.0000	-26.7715	-0.0036	-0.0128
	0.2	0.6626	2.0637	-0.0036	-0.0129

Table F.17: Grey relational co-efficient for aluminum alloy, silicon carbide, alumina and CNT

Specimen 4- Al –alloy + CNT + Silicon carbide + Alumina					
RPM	Feed rate, mm/rev	In diameter (Top), mm	Out diameter (Bottom), (mm)	Taper angle	Delamination
400	0.1	0.3333	1.0000	-0.0029	-0.0099
	0.15	0.4271	3.1583	-0.0029	-0.0099
	0.2	0.5971	1.0000	-0.0025	-0.0098
1140	0.1	0.4271	1.0000	-0.0028	-0.0099
	0.15	0.4271	1.0000	-0.0028	-0.0099
	0.2	0.5971	3.1583	-0.0028	-0.0098
2260	0.1	0.5971	1.0000	-0.0025	-0.0098
	0.15	0.4977	1.0000	-0.0026	-0.0098
	0.2	1.0000	3.1583	-0.0025	-0.0097

Step 4: Generate the grey relational grade

The grey relational grade for the number of performance characteristics can be determined using following equation.

$$\overline{\gamma_j} = \frac{1}{k}\sum_{i=1}^{m} \gamma_{ij} \tag{F.8}$$

Where $\overline{\gamma_j}$ is the grey relation grade for the jth experiment and k is the number of performance characteristics. The grading developed considering the higher the better.

Table F.18: Grey relational grade for aluminum alloy

Specimen 1- Al –alloy			
RPM	Feed rate, mm/rev	Grey relation grade	Ranking
	0.1	0.5000	7
400	0.15	0.3394	9
	0.2	0.3581	8
	0.1	0.5556	5
1140	0.15	0.5556	5
	0.2	0.8333	3
	0.1	0.8909	1
2260	0.15	0.8909	1
	0.2	0.8333	3

Table F.19: Grey relational grade for aluminum alloy and CNT composites

Specimen 2- Al –alloy + CNT			
RPM	Feed rate, mm/rev	Grey relation grade	Ranking
	0.1	-0.1569	5
400	0.15	-13.3624	9
	0.2	-0.0367	4
	0.1	0.0098	2
1140	0.15	0.0098	2
	0.2	-13.3211	8

Specimen 2- Al –alloy + CNT			
RPM	Feed rate, mm/rev	Grey relation grade	Ranking
	0.1	0.3726	1
2260	0.15	-13.1957	6
	0.2	-13.2795	7

Table F.20: Grey relational grade for aluminum alloy, silicon carbide and CNT composites

Specimen 3- Al –alloy + CNT + Silicon carbide			
RPM	Feed rate, mm/rev	Grey relation grade	Ranking
	0.1	-6.6139	9
400	0.15	0.6775	2
	0.2	0.3701	4
	0.1	-6.4470	5
1140	0.15	-6.5314	8
	0.2	-6.4470	5
	0.1	0.7619	1
2260	0.15	-6.4470	5
	0.2	0.6775	2

Table F.21: Grey relational grade for aluminum alloy, silicon carbide, alumina and CNT composites

Specimen 4- Al –alloy + CNT + Silicon carbide + alumina			
RPM	Feed rate, mm/rev	Grey relation grade	Ranking
	0.1	0.3301	9
400	0.15	0.8931	3
	0.2	0.3962	5
	0.1	0.3536	8
1140	0.15	0.8931	3
	0.2	0.9357	2
	0.1	0.3962	5
2260	0.15	0.3713	7
	0.2	1.0365	1

FSC
www.fsc.org
MIX
Papier | Fördert
gute Waldnutzung
FSC® C083411

Kolibri 360 GmbH
Ferdinand-Jühlke-Straße 7
99095 Erfurt, Deutschland
produktsicherheit@kolibri360.de

Druck:
CPI Druckdienstleistungen GmbH
im Auftrag der
Kolibri 360 GmbH
Ferdinand-Jühlke-Str. 7
99095 Erfurt